Other books by Howard W. Eves

ELEMENTARY MATRIX THEORY

FUNCTIONS OF A COMPLEX VARIABLE, VOL. 1 AND 2

FUNDAMENTALS OF GEOMETRY

AN INTRODUCTION TO THE HISTORY OF MATHEMATICS

A SURVEY OF GEOMETRY, VOL. 1 AND 2

INTRODUCTION TO COLLEGE MATHEMATICS,
coauthor with C. V. Newsom

AN INTRODUCTION TO THE FOUNDATIONS AND
FUNDAMENTAL CONCEPTS OF MATHEMATICS,
coauthor with C. V. Newsom

THE OTTO DUNKEL MEMORIAL PROBLEM BOOK,
editor with E. P. Starke 1

IN MATHEMATICAL CIRCLES, VOL. 1 AND 2

MATHEMATICAL CIRCLES REVISTED

MATHEMATICAL CIRCLES SQUARED

Translations

INITIATION TO COMBINATORIAL TOPOLOGY,
by Maurice Fréchet and Ky Fan

INTRODUCTION TO THE GEOMETRY OF COMPLEX NUMBERS,
by Roland Deaux

MATHEMATICAL

CIRCLES

ADIEU

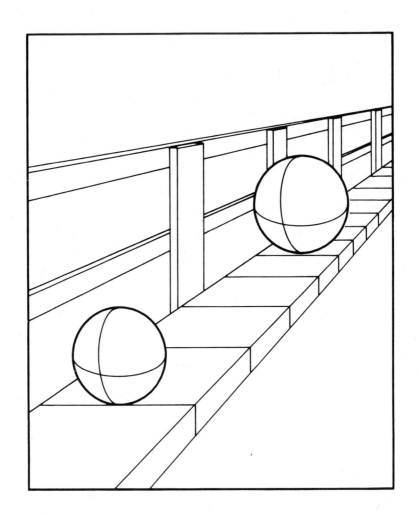

MATHEMATICAL
CIRCLES ADIEU

A FOURTH COLLECTION OF
MATHEMATICAL STORIES AND ANECDOTES

HOWARD W. EVES

PRINDLE, WEBER & SCHMIDT, INC.

Boston, Massachusetts

FRONTISPIECE: The optical illusion of perspective is mentioned in Item 1456°.

© Copyright 1977 by Prindle, Weber & Schmidt, Incorporated
20 Newbury Street, Boston, Massachusetts 02116

Library of Congress Cataloging in Publication Data

Main entry under title:

Mathematical circles adieu.

   Sequel to Mathematical circles squared, entered under H. W. Eves.
   Includes index.
   1. Mathematics—Miscellanea. I. Eves, Howard Whitley
QA99.M36   510'.2   77–22821
ISBN 0–87150–240–2

Printed in the United States of America

TO THE ADMINISTRATION,
    FACULTY AND STUDENTS

    of the University of Maine at Machias
    with fond memories and warmest wishes

PREFACE

Upon completion of the third trip around the Mathematical Circle (*Mathematical Circles Squared*), I fully intended not to make any more of these circuits. In the intervening five years, however, so many of the mathematical brethren sent me favorite anecdotes that they hoped would appear in a subsequent ramble, that I finally felt compelled to go around just once more. Herewith is the result— which I *insist* is my farewell to these pleasant wanderings.

Again I point out that many anecdotes cannot survive a careful examination for veracity; they are to be regarded more as folklore than as truth. Someone once insightfully defined an *anecdote* as "a revealing account of an incident that never occurred in the life of some famous person."

So, many thanks to all those who have supplied me with material and to those who have written kind and encouraging words about my circular travels. Special thanks go to *The Mathematics Teacher* and to *The American Mathematical Monthly,* to my friend Elwood Ede and to the good folks at the University of Maine at Machias (where I assembled the work while assisting there in the mathematics department), and to Prindle, Weber and Schmidt for their generous and sensitive understanding. Finally, I hope this fourth set of anecdotes will prolong our fond memories of two especially lovable and talented mathematicians, L. J. Mordell and Leo Moser.

Adieu, then, and perhaps at some future time we devotees shall meet in some mathematical Valhalla and enjoy ourselves by swapping more stories of the old days on earth.

HOWARD W. EVES

ix

CONTENTS

CONTENTS

Contents

CONTENTS

QUADRANT TWO

CONTENTS

CONTENTS

QUADRANT THREE

CONTENTS

CONTENTS

CONTENTS

QUADRANT FOUR

CONTENTS

CONTENTS

CONTENTS

L'ENVOI

GEOMETRICAL ILLUSIONS

CONTENTS

QUADRANT ONE

From a unique thesis
to Lewis Carroll's fireplace

MATHEMATICS IN EARLY AMERICA

MATHEMATICS was slow getting started in the early American colonies, and that which was done and taught shows a stark poverty compared with the mathematics of the same period in Europe. This was to be expected, for the time and energy of the early colonists were largely consumed by more mundane, immediate, and pressing matters. But, slow and meager as the start was, once the nineteenth century was past, mathematics in our country developed at a prodigious and ever-increasing rate, and today mathematics in the United States is second to that of no other country of the world.

For an excellent brief survey of the history of early American mathematics, see D. E. Smith and Jekuthiel Ginsburg, *A History of Mathematics in America Before 1900* (Carus Mathematical Monograph Number Five). Many of the items of the present section were suggested by this fine little book.

1° *A unique thesis.* The early American colleges were designed primarily for the training of clergy, and consequently followed the English plan of confining their work principally to Latin and Greek, with the result that courses in science, even for the master's degree, were quite trivial. The only mathematical master's thesis offered in the United States before 1700 was: "Is the quadrature of a circle possible?" in which the candidate took the affirmative position. This was in 1693 at Harvard College.

2° *The poverty of early American collegiate mathematics.* The low standards of early American collegiate mathematics are exemplified in the following titles of theses for the bachelor's degree at Yale in 1718.

1. Multiplication by a decimal fraction decreases the value of any given number; division increases it. [Apparently the case of a proper common fraction was not considered as analogous.]

2. What is involved in involution is resolved by evolution.

3

3. Given the base and altitude, the angle at the base cannot be found by the use of a line of sines. [Here we have a right triangle standing on one of its legs as base. It seems that the writer was unaware of the relation $\tan A = \sin A/(1 - \sin^2 A)^{1/2}$.]

4. Trigonometric problems can be solved most accurately by the use of logarithms.

5. The surface of a sphere is four times the area of its largest circle.

6. The angle at the base of the horizontal sundial must agree with the elevation of the pole.

3° *Low-grade work.* The low grade of mathematical work in the colleges of early America is illustrated by *The Statutes of Columbia College in the City of New York,* printed in 1785. For entrance, a candidate had merely to know the four fundamental operations of arithmetic, along with the rule of three. In the freshman year the student devoted three class sessions a week to vulgar and decimal fractions, extraction of roots, and algebra as far as quadratic equations. In the sophomore year the mathematics class met once a day to study Euclid's *Elements,* spherical trigonometry, conic sections, and the "higher branches of algebra." No further mathematics was required except that taught incidentally in the physics and astronomy courses of the junior year.

4° *Early American theologian–mathematicians.* The Puritan hope that a Holy Commonwealth would develop in America, and the zeal of such religious leaders as Cotton Mather, Jonathon Edwards, George Whitefield, and John Woolman, naturally left a mark on the early American colleges. In a similar way, Samuel Johnson, the first president of King's College, from which Columbia University later evolved, and Dean (later to become Bishop) George Berkeley tried to revert the people of the colonies back to the Anglican Church. Berkeley later wrote diatribes against the weaknesses in the foundations of Newton's theory of fluxions, and his *Collected Works* deal extensively with mathematics. Dean Berkeley came to Newport, Rhode Island, in 1729 and remained in America for three years.

4

It is because of this early religious zeal that the first colleges in America, like those in England and Ireland, were established primarily to train the clergy. Thus, up to about 1850, a large proportion of the professors of mathematics in America were clergymen with more interest in theology than in mathematics.

5° *George Berkeley in America.* George Berkeley (1685–1753), who became a well-known Anglican bishop and a severe critic of Newton's theory of fluxions, earlier in life entertained the idea of establishing a college in Bermuda for the education of young colonists and Indians of the American mainland. Though he gathered some high support for his plan, the project finally had to be dropped, but not before he spent three years (1729–1731) in Newport, Rhode Island, where he gave considerable encouragement to higher education in the American colonies. He is still honored at Harvard and at Yale, and Berkeley, California, was named after him.

6° *The Winthrop mathematics library.* Important libraries, from the historical point of view, sometimes settle in unexpected places and rest there often unknown even to most of the local scholars. Perhaps the best evidence of the mathematical material available in America in the mid-eighteenth century is to be found in a collection of over a hundred old texts now residing in the library of Allegheny College in Meadville, Pennsylvania. It was the mathematics library of John Winthrop (1714–1779), one of four Winthrops bearing that name, and a descendent of the first governor of the Massachusetts Bay Colony.

John Winthrop was one of the most learned men in America of his time. When only twenty-four, he was appointed to the Hollis professorship at Harvard, a post that he held for over forty years. Among his first self-appointed tasks at Harvard was to master Newton's *Principia* and to make full use of the astronomical instruments that Thomas Hollis had earlier had sent to Harvard from England. Among the instruments was a telescope which had belonged to Edmund Halley (1656–1742). Winthrop was an astronomer rather than a mathematician, and the original records of his astronomical observations of 1739 are still in the Harvard library. It was in 1740 that he made the first observation in America of a transit of the planet Mercury across the face of the sun, and later, in 1761, he traveled to Newfoundland to observe another such transit, this trip probably being the first purely scientific expedition sent out by an American colony. In 1766 he was elected a Fellow of the Royal Society. Two years later, in 1768, he was awarded the degree LL.D. by the University of Edinburgh and was chosen a member of the American Philosophical Society. He was the first to receive the degree LL.D. from Harvard (in 1773) and he was the principal founder of the American Academy of Arts and Sciences.

Winthrop's above-mentioned mathematics library, now housed at Allegheny College, contains over one hundred books on pure and applied mathematics, and almost all of them are historical classics. The collection contains works by Barrow, Cassini, Cotes, De Lalande, Desaguliers, Descartes, Euclid, Gravesande, Halley, Huygens, Keill, Maclaurin, Maseres, Newton, Oughtred, Ramus, Whiston, and Wolf. Many of these books are very rare, and any one of them would occupy a worthy place in any mathematical collection. One would think this collection would have settled at Harvard. There must be a story as to how Allegheny College acquired the volumes. When the writer of these anecdotes spent a week lecturing at Allegheny College a few years ago, none of the members of the mathematics staff there was aware of the presence of the Winthrop collection.

A collection of over 150 mathematical classics that were in the Harvard library in the first quarter of the eighteenth century was completely consumed by fire in 1764.

7° *Jefferson on Rittenhouse.* David Rittenhouse (1732–1796) was, after John Winthrop, the next noted early American astronomer with an interest in applied mathematics. Starting as a watchmaker, he became a constructer of precision scientific instruments. As an able practical mathematician, he early worked as a surveyor and played a part in the establishment of the Mason-Dixon line. He made a remarkably precise observation of the transit of Venus on June 3, 1769, and very accurately calculated the elements of the future transit of Venus of December 8, 1874. The high quality of his papers in the first four volumes of the *Transactions of the American Philosophical Society* led to his election as president, in 1790, of the Society, succeeding Benjamin Franklin. He was a Fellow of the American Society of Arts and Letters and of the Royal Society of London, and in 1789 was awarded the degree LL.D. by Princeton University. He served as vice-provost at the University of Pennsylvania and was among the first to use spider lines in a telescope. In 1792, George Washington appointed him Director of the Mint, a position he held until the year before his death, and one is reminded of the last years of Isaac Newton.

Thomas Jefferson, always ready to defend his country against critics and to extol the achievements of his countrymen, wrote the following extravagant lines about Rittenhouse and his construction of a planetarium:

> We have supposed Rittenhouse second to no astronomer living; that in genius he must be first because he is self taught. As an artist he has exhibited as great a proof of mechanical genius as the world has ever produced. He has not indeed made a world; but he has by imitation nearer approached its Maker than any man who has lived from Creation to this day.

8° *Burning the calculus.* It was a common custom among mathematics students in American colleges in the late nineteenth century ceremoniously to burn their calculus texts at the close of the sophomore year, thus displaying by a great bonfire their estimate of the value of the subject as then taught. [In this connection, see Item 139° of *Mathematical Circles Revisited.*]

9° *How Charles Gill became a mathematician.* There are a number of names that were prominent in American mathematics in the first half of the nineteenth century but which are now almost completely forgotten—names like Robert Adrain (1775–1843), Robert Patterson (1743–1824), William Rogers (1804–1882), John Farrar (1779–1854), Theodore Strong (1790–1869), Alexander Dallas Bache (1806–1867), and Ferdinand Rudolph Hassler (1770–1843).

Among these earlier and now almost forgotten men was Charles Gill, who had an interesting and unusual entry into mathematics. Gill was born in Yorkshire, England, in 1805, the son of a local shoemaker. When thirteen, Gill went to sea and voyaged to the West Indies. Three years later, when he was only sixteen, the rapid and tragic loss by yellow fever of all the officers of the ship on which he was serving, catapulted him into command of the vessel. His navigational skill brought the ship safely to its destined port, and he became convinced that mathematics was his forte. He accordingly gave up the sea and began a career of teaching mathematics, spending his leisure time studying the subject. He contributed to a number of mathematical journals and was already an established thinker when he came to America in 1830.

In this country, Gill immediately engaged in teaching mathematics in secondary schools, his final position being in the Flushing Institute on Long Island, N.Y. It was while at this school, which grew into a short-lived college, that Gill brought out the *Mathematical Miscellany,* a semi-annual periodical devoted largely to the proposal and solution of fairly difficult and advanced mathematics problems. The journal, though it lasted only through the four years 1836–1839, exerted a considerable influence on American mathematics. Gill himself possessed real skill in solving intricate problems in number theory. Solutions by him of such problems appeared in such popular journals as *The Ladies' Diary, The Gentleman's Mathematical Companion, The Educational Times,* and his own *Mathematical Miscellany.* He was also interested in actuarial mathematics, and has been called "the first actuary in America." Gill is mentioned several times in Dickson's *History of the Theory of Numbers.*

10° *A "paradoxical" relation.* Benjamin Peirce (1809–1880) was generally regarded as the leading mathematician in America of his time. He graduated from Harvard in 1829 when only twenty years old, became a tutor there in 1831, then soon after, in 1833, was awarded the professorship in mathematics and natural science, and still later, in 1842, the professorship in mathematics and astronomy.

The most often told anecdote about Benjamin Peirce concerns an incident that occurred in one of his classes on function theory. Having established the well-known and remarkable relation

$$e^{\pi/2} = {}^i\!\sqrt{i}$$

on the blackboard, he turned to the class and said, "Gentlemen, that is surely true, it is absolutely paradoxical, we cannot understand it, and we don't know what it means, but we have proved it, and therefore we know it must be the truth."

The above relation is perhaps even more striking in the equivalent form

$$e^{i\pi} = -1,$$

for here we have four of the most important numbers of mathematics, namely e, i, π, and -1, all related by a very simple expression.

11° *A molder of college presidents.*   Among Benjamin Peirce's students was Abbott Lawrence Lowell, who graduated in 1877 at the age of twenty-one with highest honors in mathematics. Lowell's paper on "Surfaces of the second order as treated by quaternions" attests to his continued interest in mathematics. In 1909 Lowell became president of Harvard University, serving in that capacity for twenty-four years.

Thomas Hill (1818–1891) was another and earlier student of Peirce, but Hill did not achieve the level of mathematical recognition later accorded to Lowell. Hill became president of Harvard University in 1863.

12° *The Peirce family.*   The Peirce family was one of the most prominent families in early American mathematics. We have already noted (in Item 10°) that Benjamin Peirce (1809–1880) was a famous professor of mathematics at Harvard and was considered the leading American mathematician of his time. Benjamin Peirce had two sons, James Mills Peirce (1834–1906) and Charles Sanders Peirce (1839–1914). James Mills Peirce served as an assistant professor of mathematics at Harvard for the eight years from 1860 to 1869, and as a professor of mathematics from 1869 to 1906. He

9

was largely an applied mathematician. Charles Sanders Peirce was perhaps the greatest mathematician of the Peirce family, though it is only in more recent times that much of his work has been properly appreciated. He graduated from Harvard in 1859 at the age of twenty, and from the Lawrence Scientific School of Harvard in 1863. For a number of years he served on the United States Coast and Geodetic Survey and achieved a fame for his work on geodesy. In 1880 he lectured at Johns Hopkins University on philosophical logic. The latter part of his life was devoted chiefly to mathematical logic; it is this work that has only rather recently been recognized as truly trail-blazing. Another mathematical member of the Peirce family was Benjamin Osgood Peirce (1854–1914), a second cousin once removed of Benjamin Peirce. This Peirce wrote valuable papers in mathematical physics. His *Mathematical and Physical Papers 1903–1913* was published by the Harvard Press in 1926.

13° *Appreciation from a master.*   George William Hill (1838–1916), the third president of the American Mathematical Society, ranks as one of the better known astronomers who contributed noteworthily to pure mathematics. He graduated from Rutgers University in 1859 and shortly thereafter joined the staff of the *Nautical Almanac.* In astronomy he devoted most of his energies to lunar theory; in mathematics he concentrated on differential equations, determinants, and series, and introduced infinite determinants into mathematics. It was Hill's memoir on infinite determinants that some nine years later led Henri Poincaré to take up the study of their convergence. Other work of Hill lay at the basis of Poincaré's *Les méthodes nouvelles de la méchanique.*

When Professor Robert Simpson Woodward (1849–1924), one of the early prominent supporters of the American Mathematical Society, and its fifth president, introduced Hill to Poincaré on the latter's visit to America, Poincaré's first words as he gripped Hill's hand were, "You are the one man I came to America to see."

14° *A descendent of Benjamin Franklin.*   Benjamin Franklin was an amateur dabbler in magic squares and a few other things of a mathematical nature (see Item 317° of *In Mathematical Circles*).

One might fairly wonder if there ever was a descendent of Benjamin Franklin who might more properly be classified as a mathematician. There was—Alexander Dallas Bache (1806–1867). Though no truly outstanding figure in mathematics, Bache did author several articles on astronomy and surveying in the *Proceedings* and the *Transactions of the American Philosophical Society*. He was, from 1828 to 1841, a professor of natural philosophy and chemistry at the University of Pennsylvania. In 1843 he became superintendent of the United States Coast Survey, and it was here that he made his reputation.

PIERRE DE FERMAT AND RENÉ DESCARTES

WE have noted, in the previous section, the paucity of mathematics and the complete lack of top-flight mathematicians in America in the seventeenth century. In sharp contrast to this sterile picture, mathematics in Europe was thriving, and indeed was experiencing a remarkable period of growth.

It was early in the seventeenth century that Napier revealed his invention of logarithms, Harriot and Oughtred contributed to the notation and codification of algebra, Galileo founded the science of dynamics, and Kepler announced his laws of planetary motion. Later in the century, Desargues and Pascal opened up the field of projective geometry, Fermat fathered modern number theory, Fermat and Descartes gave us analytic geometry, and Fermat, Pascal, and Huygens laid the foundations of the modern theory of mathematical probability. Then, toward the end of the century, after a host of seventeenth-century mathematicians had prepared the way, the epoch-making creation of the calculus was made by Newton and Leibniz. The seventeenth century was truly a remarkable period for mathematics in western Europe.

Of the great European men of mathematics mentioned above, we now tell a few stories about the two Frenchmen, Fermat and Descartes. For many other stories and anecdotes about these men, and about the others mentioned above, the reader may consult our previous three trips around the mathematical circle.

15° *The enigma of Fermat's birthdate.*   There is a seemingly reliable report that has come down to us asserting that Fermat was born at Beaumont de Lomagne, near Toulouse, on August 17, 1601. It is known that Fermat died at Castres or Toulouse on January 12, 1665. His tombstone, originally in the church of the Augustines in Toulouse and then later moved to the local museum, gives the above date of death and Fermat's age at death as fifty-seven years. Because of the above conflicting data, Fermat's dates are usually listed as (1601?–1665). Indeed, for various reasons, Fermat's birth year, as given by different writers, ranges from 1590 to 1608.

16° *How did Fermat find the time?*   Fermat did so much top-quality mathematics, in addition to serving in the local parliament of Toulouse, that one naturally wonders where he found the time for all his creative work. It has been suggested that Fermat's position as a *King's councillor* in the parliament of Toulouse is perhaps the key. Unlike many other public servants, a King's councillor was expected to remain aloof from his fellow townsmen and to abstain from most social activities, so that he might less easily be corrupted by bribery and might the better carry out his duties without the stress of partisan influence. In short, Fermat's position gave him plenty of leisure time.

17° *The sixth Fermat number.*   Fermat conjectured that $p_n = 2^{2^n} + 1$ is a prime for all nonnegative integers n. Now for $n = 0$, 1, 2, 3, and 4 we find $p_n = 3, 4, 17, 257$, and $65,537$, respectively, all of which are prime numbers. On the other hand, Euler, a century later, showed that $p_5 = 4,294,967,297$ contains 641 as a factor, and therefore is not a prime number. Thus Fermat's conjecture has been proven false. The American lightning calculator, Zerah Colburn, when a mere boy, was once asked if $4,294,967,297$ is prime or not. After a short mental calculation he asserted that it is not, as it has the divisor 641. He was unable to explain the process by which he had reached his amazing correct conclusion.

For further anecdotal information concerning the Fermat numbers, see Item 178° of *In Mathematical Circles*.

18° *Fermat and his method of infinite descent.* In a letter of August, 1659, to Carcavi, Fermat gave the following clear and concise account of his method of infinite descent. We give here a rather free translation of the pertinent part of the letter.

"For a long time I was not able to apply my method to affirmative propositions, because the twist and the trick for getting at them is much more troublesome than that which I use for negative propositions. Thus, when I had to prove that *every prime of the form* $4n + 1$ *is the sum of two squares,* I found myself in a fine torment. But at last a meditation many times repeated gave me the light I lacked, and now affirmative propositions submit to my method, with the aid of certain new principles which necessarily must be adjoined to it. The course of my reasoning in affirmative propositions is such: if an arbitrary chosen prime of the form $4n + 1$ is not a sum of two squares, there will be another of the same nature, less than the one chosen, and next a third still less, and so on. Making an infinite descent in this way we finally arrive at the number 5, the least of all numbers of this kind. It follows that 5 is not the sum of two squares. But it is. Therefore we must infer by a *reductio ad absurdum* that all prime numbers of the form $4n + 1$ are sums of two squares."

Of course, the hard part of applying Fermat's method of infinite descent lies in the very first step, namely proving that if the assumed conjecture is true of any number of the concerned kind, chosen at random, then it will be true of a smaller number of the same concerned kind. In connection with the problem that any prime of the form $4n + 1$ is a sum of two squares, Fermat never recorded for posterity just how he carried out the first step of his method of infinite descent. The result was finally proved by Euler in 1749 after he had struggled, off and on, for seven years to find a proof.

E. T. Bell has pointed out that proofs of many theorems in number theory, like the one above, are so evasive that "it requires more innate intellectual capacity to dispose of the apparently childish thing than it does to grasp the theory of relativity." In the same connection, Gauss wrote: "A great part of its [the higher arithmetic] theories derives an additional charm from the

peculiarity that important propositions, with the impress of sim-
plicity on them, are often easily discovered by induction, and yet
are of so profound a character that we cannot find the demonstra-
tions till after many vain attempts; and even then, when we do
succeed, it is often by some tedious and artificial process, while the
simple methods may long remain concealed."

For a simple illustration of Fermat's method of infinite descent, proving that $\sqrt{2}$ is irrational, see Item 179° of *In Mathematical Circles.*

19° *Descartes versus Fermat.*   Descartes was frequently irrita-
ble, quick-tempered, and acid in his mathematical controversies
with Fermat. Fermat, on the other hand, was always patient, even-
tempered, and unaffectedly courteous. These differing qualities in
the two men are reflected by their avocations—Descartes was a
soldier and Fermat was a jurist.

20° *The Descartes miracle.*   Perhaps the most remarkable
thing about Descartes' life is that it was not brought to a sudden
and premature end by a rapier or a musketball. Descartes spent a
large part of his life soldiering, and though he served much inac-
tive time in army camps, he also on a number of occasions found
himself in the fray of battle. Thus he could easily have lost his life
in 1620, at the young age of 24, when, enlisted under the Elector
of Bavaria, he engaged in the very real fighting of the battle of
Prague. Later he underwent a bloody piece of soldiering, with
distinction, under the Duke of Savoy. Still later, under the King of
France, Descartes' life could have terminated at the siege of La
Rochelle. And if soldiering was not enough, consider the time
when, without his usual complement of bodyguards, he took a boat
for east Frisia and was set upon by a cutthroat crew that sought to
rob him and then throw his body overboard; he whipped out his
sword and compelled the crew to row him back to shore. The
Descartes miracle is that analytic geometry managed to escape the
accidents of battle, sudden death, and murder.

21° *The cavalier.*   Once a half-drunken lout insulted Descartes' lady of the evening. In true cavalier fashion, Descartes went after the sot in stirring D'Artagnan manner and soon flicked the sword out of the fool's hand. He spared the sot's life since he felt it would be too messy to butcher the man in front of a beautiful lady.

22° *Descartes' daughter.*   Though Descartes never married, he did have a daughter by one of his lady friends. The child died early and Descartes was deeply affected.

23° *Overcoming hypochondria.*   Descartes' delicate childhood infected him with hypochondria, and for years he had an oppressive dread of death. Then, at middle age, he came to the conclusion that nature is one's best physician and that the secret of keeping well is to shun the fear of death. Thus he overcame his hypochondria.

24° *Descartes' bones.*   The death of Descartes in Stockholm, while attempting to bring learning to the court of the young and willful Queen Christina of Sweden, has been told in Item 177° of *In Mathematical Circles*. The great philosopher–mathematician was entombed in Sweden and efforts to have his remains transported to France failed. Then, seventeen years after Descartes' death, when Christina had long cast off her crown and her faith, the bones of Descartes, except for those of his right hand, were returned to France and re-entombed in Paris at what is now the Panthéon. The bones of the right hand were secured, as a souvenir, by the French Treasurer-General for his skill in engineering the transportation.

Commenting on the return of Descartes' remains to his native land, Jacobi remarked, "It is often more convenient to possess the ashes of great men than to possess the men themselves during their lifetime."

15

SOME PRE-NINETEENTH-CENTURY
MATHEMATICIANS

HERE are a few random anecdotes, not already told in our earlier trips around the mathematical circle, ranging from Greek antiquity up through the eighteenth century.

25° *A voluptuous moment.* Everyone knows the story of Hobbes's first contact with Euclid: opening the book, by chance, at the theorem of Pythagoras, he exclaimed, "By God, this is impossible," and proceeded to read the proofs backwards until, reaching the axioms, he became convinced. No one can doubt that this was for him a voluptuous moment, unsullied by the thought of the utility of geometry in measuring fields.

—BERTRAND RUSSELL
In Praise of Idleness. London: George Allen and Unwin, Ltd.,
1935.

26° *Faith versus reason.* . . . There can be no doubt about faith and not reason being the *ultima ratio.*

Even Euclid, who has laid himself as little open to the charge of credulity as any writer who ever lived, cannot get beyond this. He has no demonstrable first premise. He requires postulates and axioms which transcend demonstration, and without which he can do nothing. His superstructure indeed is demonstration, but his ground is faith. Nor again can he get further than telling a man he is a fool if he persists in differing from him. He says "which is absurd," and declines to discuss the matter further. Faith and authority, therefore, prove to be as necessary for him as for anyone else.

—SAMUEL BUTLER
From Chapter Sixty-five of *The Way of All Flesh.*

27° *The Gresham chair in geometry.* The earliest professorship of mathematics established in Great Britain was a chair in geometry founded by Sir Thomas Gresham (1519?–1579) at

Gresham College in London. Henry Briggs, who was the first occupant of the Savilian chair at Oxford, also had the honor of being the first to occupy the Gresham chair in geometry.

28° *A tenuous connection with mathematics.* Sir Thomas Gresham, known in mathematics as the founder of the first professorship of mathematics in Great Britain, namely the Gresham chair in geometry at Gresham College of London, also founded England's Royal Exchange. Accordingly, when a weather vane was constructed for Faneuil Hall, the famous trading hall of colonial Boston, the designer of the vane fashioned it in the form of a copper grasshopper, for this familiar insect appears on the crest of Sir Thomas Gresham. The designer was Shem Drowne, a metalsmith of Kittery, Maine, who late in the seventeenth century moved to Boston.

The famous grasshopper weather vane atop of Faneuil Hall was noticed missing on January 5, 1974, and there was much speculation about the possibility of thieves having acquired the valuable vane by means of a swooping helicopter. The vane, valued in the hundreds of thousands of dollars, was later prosaically rediscovered wrapped in rags where would-be thieves had left it in the Faneuil Hall belfry.

Four of Shem Drowne's weather vanes are still surviving, and three of these are still in use: the North Church banner (that capped the Old North Church at the time signal lanterns were hung there to warn that the British were moving toward Concord by sea); a rooster vane on a church in Cambridge; and the priceless grasshopper vane of Faneuil Hall.

29° *Gresham's law.* Gresham's law, named after Sir Thomas Gresham, founder of the Royal Exchange of London, states that "bad money tends to drive good money out of circulation." Gresham stated the principle as follows: "When by legal enactment a government assigns the same nominal value to two or more forms of circulating medium whose intrinsic values differ, payments will always, as far as possible, be made in that medium of which the cost of production is least, the more valuable medium tending to disappear from circulation."

Thus, in America in 1896, gold coins were hoarded because of the chance that the government might make silver coins and proclaim them to be legally of the same value as the gold coins. Again, in Europe just after World War I, so much paper money was made that the intrinsically more valuable metal money all but disappeared. Indeed, when paper money was made in America, possessers of silver dollars hoarded them as more valuable than the legally equivalent dollar bills.

As early as the thirteenth century, dishonest dealers would shave the edges of gold coins, and those who received these lighter pieces usually passed them along as quickly as possible, retaining the full-weighted coins as long as they could.

30° *The Savilian and Lucasian professorships.*   Many distinguished British mathematicians have held either a Savilian professorship at Oxford or a Lucasian professorship at Cambridge.

Sir Henry Savile was one time warden at Merton College at Oxford, later provost of Eton, and a lecturer on Euclid at Oxford. In 1619, he founded professorial chairs at Oxford, one in geometry and one in astronomy. Henry Briggs (of logarithm fame) was the first occupant of the Savilian chair of geometry at Oxford. John Wallis, Edmund Halley, and Sir Christopher Wren are other seventeenth-century incumbents of Savilian professorships.

Henry Lucas, who represented Cambridge in parliament in 1639–1640, willed the university resources for the founding in 1663 of the professorship that bears his name. Isaac Barrow was elected the first occupant of this chair in 1664, and six years later was succeeded by Isaac Newton.

31° *Galileo's power.*   The two things most universally desired are power and admiration. Ignorant men can, as a rule, only achieve either by brutal means, involving the acquisition of physical mastery. Culture gives a man less harmful forms of power and more deserving ways of making himself admired. Galileo did more than any monarch has done to change the world, and his power immeasurably exceeded that of his persecutors. He had therefore no need to aim at becoming a persecutor in his turn.

—BERTRAND RUSSELL

In Praise of Idleness. London: George Allen and Unwin, Ltd., 1935.

32° *A clerihew.* A *clerihew* is a form of light verse, akin to the limerick, that became popular in England. It is named after its inventor, Edmund Clerihew Bentley, well known as the author of *Trent's Last Case,* and the friend to whom G. K. Chesterton dedicated his entertaining fantasy, *The Man Who Was Thursday.* Bentley's best known clerihew concerns the mathematician–architect Sir Christopher Wren (1632–1723):

> Sir Christopher Wren
> Said, "I am going to dine with some men.
> If anybody calls
> Say I'm designing St. Paul's."

As another example of a mathematical clerihew, there is the following, given by J. C. W. de La Bere in the December 1974 issue of the *Australian Mathematical Society Gazette*:

> Archimedes of Syracuse,
> To get into the news,
> Called out "Eureka"
> And became the first streaker.

33° *Euler and the seat of honor.* In 1783 Czarina Katherine II appointed the talented and celebrated Princess Daschkoff to the directorship of the Imperial Academy of Sciences in Petersburg. Since women were seldom so highly honored in those days, the appointment received wide publicity.

The Princess decided to commence her directorship with a short address to the assembled savants of the Academy. She invited Euler, then elderly and blind but the most respected scientist in Russia, to be her special guest of honor. Euler, with a son and a grandson, accompanied the Princess to the Academy in her personal coach. In her brief address, the Princess stressed her high respect for the sciences and said she hoped that Euler's presence would serve as a source of inspiration to the entire Academy. She then sat down, intending Euler to occupy the chair of honor next to her, but before Euler could be led to the seat, Professor Schtelinn dropped himself into the honored place. Some years ago, Schtelinn had been recognized by Peter III, had served as a councilor of state, and had held the rank of Major-General, and he now felt that he deserved first place in the Academy body.

When Princess Daschkoff saw Schtelinn settling himself next to her, she turned to Euler and said, "Please be seated anywhere, and the chair you choose will naturally be the seat of honor." This act charmed Euler and all present—except the arrogant Professor Schtelinn.

34° *A curious error.*   Leonhard Euler assembled a physics textbook entitled *Lettres à une Princesse d'Allemagne sur divers sujets de physique et de philosophie* [*Letters to a German Princess on various topics of physics and philosophy*]. The letters referred to were originally addressed to Princess Phillipine von Schwedt, niece of Frederick the Great. During the Seven Years War (1756–1763), Princess Phillipine and the entire Berlin court sojourned in Magdeburg. It was during this period that Euler tutored the Princess by letters written from his home in Berlin.

In the letter dated August 27, 1760, Euler enters into an explanation to his royal student of how a surveyor uses a level. He asks the Princess to imagine a straight line drawn from her quarters in Magdeburg to his home in Berlin, and then asks whether this line would be horizontal. He goes on to answer his own question in the negative, stating that the Berlin end of the line would be higher than the Magdeburg end. He explains that Berlin lies on the Spree and Magdeburg on the Elbe, and since the Spree flows into the Havel, and the Havel flows into the Elbe, Magdeburg must be nearer sea level than Berlin.

Now, actually, the Elbe at Magdeburg is 41 meters above sea level while the Spree at Berlin is only 33 meters above sea level, a relation exactly opposite to that deduced by Euler. Since the Spree does indeed flow into the Havel, and the Havel into the Elbe, wherein lies the flaw in Euler's reasoning? Consultation of a map (see our accompanying rough sketch in Figure 1) shows that, though the Havel flows into the Elbe, it does so far below Magdeburg. It is difficult to comprehend how Euler failed to take this fact into account, and possible explanations of the great mathematician's carelessness have been offered.

Some feel that Euler's error was simply that he assumed Magdeburg was near Wittenberge, which lies just below the junction of the Havel with the Elbe. If the imagined line had run from Wittenberge to Berlin, Euler's conclusion would have been correct. The fact is, however, that Magdeburg lies some 55 miles above the junction of the two rivers.

Elwood Ede, formerly of the University of Maine mathematics faculty, suggests a different and perhaps more plausible explanation. He feels that it is very likely Euler merely misjudged the direction of flow of the Elbe, assuming it runs from Wittenberge to Magdeburg instead of from Magdeburg to Wittenberge. Such a misjudgment could easily have occurred, says Ede, if Euler used a map, perhaps cropped, instead of personal observation. In this light, Ede contends, the story becomes somewhat trite and the error, though it exists, is overemphasized.

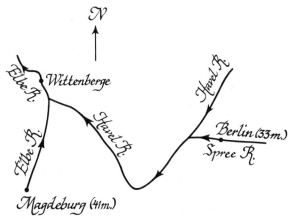

FIGURE 1

35° *Poisson's false start in life.* Siméon Denis Poisson (1781–1840), who became one of the great applied mathematicians of his time, was first destined, much against his own wishes, to be a doctor. The education was undertaken by his uncle, who started the boy off with pricking veins in cabbage leaves with a lancet. When he had perfected himself in this, he was graduated to putting on blisters. But in almost the first case in which he did this by himself, his patient died within a few hours. Although the doctors of the place assured him that "the event was a very common one," he vowed to have nothing more to do with the profession.

36° *The good life.* Poisson once remarked: "Life is good for only two things, discovering mathematics and teaching mathematics."

CARL FRIEDRICH GAUSS

A N astonishing number of anecdotes have come down to us concerning the great German mathematician Carl Friedrich Gauss (1777–1855). We have already, in our earlier trips around the mathematical circle, told over forty of them. Here are nine more.

37° *An unintended pun.* Wilhelmine Gauss, the second child of Carl Friedrich Gauss and his first wife, was named after Gauss's close friend, the noted astronomer Heinrich Wilhelm Matthias Olbers of Bremen. Gauss has told a pretty story about Minna, as Wilhelmine was always called, when she was five years old (1813). One evening he told the little girl that the small, light, rosy clouds in the sky are called "cirri", or "fleecy", clouds. A few evenings later, when the same kind of clouds reappeared, Minna tried to detain her father by saying, "Stay a little longer, Papa, the sky is so *belämmert* this evening." [*Belämmert* means "belambed" (suggested to Minna by the "fleecy" clouds), but sounds very much like *belemmert*, which means "befouled".]

38° *Minna gets married.*   In 1830 Minna Gauss married
Heinrich Ewald, the young professor of theology and Oriental
languages at Göttingen University. Ewald was incredibly absent-
minded in nonacademic matters, a trait that manifested itself in
connection with his engagement and marriage to Minna. Friends
had determined that the bachelor Ewald should, for his own good,
get married. Ewald obligingly agreed, leaving the selection of the
lady to the friends, who soon settled on Gauss's two daughters
Minna and Theresa as the best possibilities. They decided to leave
the choice between these two young ladies to Ewald himself. Accordingly, a tea was arranged at the Gauss home at which the two
daughters were to take turns pouring. Ewald was instructed to
indicate his choice by accepting tea from the daughter he pre-
ferred, a procedure to which he readily acquiesced. On the way
home after the tea, the matchmakers congratulated Ewald and
explained that Gauss was quite agreeable to Ewald wooing Minna,
from whom he had accepted tea. Ewald didn't know what his
friends were talking about, for he had completely forgotten, as
though he had never heard about it, the whole plot and even the
purpose of the tea. But he did woo Minna and married her on
September 15, 1830. Before the ceremony on the wedding day,
Ewald disappeared, and, after a frantic search, was finally found
atop a ladder in front of his bookshelves; he had forgotten all about
the impending affair. Curiously enough, the marriage turned out
to be a very happy one for both Ewald and Minna, and Gauss was
highly pleased to have his colleague Ewald as a son-in-law.

39° *Ewald babysits.*   Minna was of frail health and died of
tuberculosis on August 12, 1840, when she was only thirty-two
years of age. Some years later, her husband Ewald remarried and
in 1850 a daughter, whom Gauss regarded almost as his own
grandchild, was born of this second marriage. One day, when
Ewald's second wife had to go shopping, the baby was left in care
of her father, who was very engrossed in some Arabic grammar.
When the wife returned home, she could not find the baby, and
Ewald was equally perplexed over the child's disappearance. Then
a small whimpering revealed that the baby was in a closed dresser

drawer, where Ewald apparently had placed it, perhaps thinking it was the safest place for the child during his wife's absence.

One wonders about some of Gauss's colleagues. For two stories concerning the amazing logic of George Julius Ribbentrop, professor of law at Göttingen University, see Items 190° and 191° of *Mathematical Circles Squared*. Nor was the Prince himself free of occupational maladies—see Item 193° of *Mathematical Circles Squared*.

40° *Gauss and Ceres.*   There is a collection of over 1500 minor planets, called asteroids, almost all of whose orbits lie between those of Mars and Jupiter. The largest, and first discovered, of these is Ceres, with a diameter of about 470 miles, or about one-fifth that of the moon; the smallest asteroids have diameters under a mile.

Ceres was discovered in Palermo on New Year's day of 1801 by the Italian astronomer Giuseppe Piazzi (1746–1826), and a few observations of its positions were made during the short six weeks it was in view before it disappeared in the western sky. Astronomers awaited its reappearance in the eastern sky, but in spite of an extensive search, they were unable to find it. From the meager observations that had been made of Ceres, Gauss managed to compute the entire orbit of the little planet, with the result that on New Year's Day of 1802, Heinrich Wilhelm Matthias Olbers (1758–1840), German astronomer and physician of Bremen, was able to refind the elusive body. This mathematical calculation probably brought more popular fame to Gauss than anything else he did during his lifetime; it also commenced a long friendship between Gauss and Olbers.

It is interesting that the three principals involved in the above story were subsequently honored in celestial nomenclature, for the 1000th, 1001st, and 1002nd asteroids discovered were named Piazzia, Gaussia, and Olbersia, respectively.

41° *A priceless souvenir.*   It was at the age of nineteen that Gauss discovered that a regular polygon of seventeen sides can be constructed with straightedge and compasses, and it was this discovery, Gauss once declared, that finally settled for him that mathematics would be the field of study upon which he would spend his life and talents. The slate upon which Gauss carried out his work on the construction of regular polygons with Euclidean tools was given by him, in a moment of ecstasy and affection, to his deep personal college-day friend Wolfgang Bolyai. For the rest of his life, Bolyai kept and treasured the slate as a much-revered souvenir of his friendship with and admiration for the great mathematician. One wonders where that historic slate is today.

For the disposition of another rare and precious Gauss souvenir, see Item 202° of *Mathematical Circles Squared.*

42° *An unusual honor.* The three-masted schooner used by the 1903 German expedition to the South Pole, headed by Professor Erich von Drygalski of Munich, was named the *Gauss,* and a volcanic peak discovered by the expedition was named the *Gaussberg.*

Some years later the schooner was bought by the United States government, its name was changed, and it was put into service along the west coast of North America.

43° *Gauss's pocket watch.* Gauss ceased breathing at 1:05 A.M. on February 23, 1855. His pocket watch, which he had carried with him most of his life, ceased ticking at almost exactly the same time.

44° *Gauss's brains.* With the permission of the family, Gauss's skull and brains were carefully weighed and measured after his death. Though the brains exhibited unusually deep convolutions, they were not particularly weighty or bulky. The brains of Byron, Cuvier, and Schiller were heavier, Dante's lighter. Robert Gauss, a grandson of the great mathematician and a brilliant Colorado newspaper editor, left instructions for his brains to be weighed when he died; they were three ounces heavier than those of his famous grandfather. Gauss's brains, and also those of Dirichlet, are preserved in the department of physiology at Göttingen University.

45° *The Prince of Mathematicians.*   On January 10, 1855, by orders of the King of Hanover, the official court sculptor, Christian Heinrich Hesemann, arrived at the Gauss home to start work on a medallion of the great mathematician, who was seriously ill and not expected to live much longer. Hesemann suddenly died, on May 29, 1856, before the work was finished, and the medallion was completed by C. Dopmeyer, another sculptor of Hanover. The medallion was later placed, as a plaque, on Gauss's tombstone in the St. Albans Cemetery in Göttingen. Right after the death of Gauss on February 23, 1855, the King of Hanover ordered that a commemorative medal be prepared in honor of Gauss. This seventy-millimeter medal was in time (1877) completed by the well-known sculptor and medalist, Friedrich Brehmer, of Hanover, and was based on the earlier medallion. On it appears the inscription:

> Georgius V. rex Hannoverge
> Mathematicorum principi
> [George V, King of Hanover
> to the Prince of Mathematicians]

Ever since, Gauss has been known as "the Prince of Mathematicians."

SCHELLBACH AND GRASSMANN

KARL Schellbach (1804–1892) [Cajori says he was born in 1809] and Hermann Günther Grassmann (1809–1877) each spent a large part of his life teaching mathematics to younger students, and each possessed deep religious feelings. Schellbach, though enormously admired and of considerable fame in his day, is little known in mathematical circles now, for his forte was the teaching of mathematics rather than the creation of it. On the other hand, Grassmann, though unnoted and obscure in his day, is widely known in mathematical circles now, for his forte was the creation of mathematics rather than the teaching of it. It is thus that history awards its laurels.

A number of pretty stories about Schellbach and Grassmann have been preserved for us by Dr. W. Ahrens in his engaging little booklet *Mathematiker-Anekdoten.* We revive some of these stories here, adapting free translations from Ahrens made by Mr. Elwood Ede.

46° *Preaching the gospel of mathematics.*   Karl Schellbach, the romantic poet and famous mathematics professor at the Friedrich Wilhelm Gymnasium in Berlin, considered the teaching of mathematics to be a religious vocation, and he believed that mathematicians were priests who should expose as many people as possible to the realms of mathematical blessedness and glory. He felt that both the gifted and the intellectually poor should take part in this kingdom of heaven. The Minister of Culture, von Bethmann-Hollweg, grandfather of the fifth Reichskanzler, once commented that Schellbach's teaching was "an inspired hymn of mathematics." In an effort to expose inexperienced teachers to the mathematical instruction of Schellbach, the ministry introduced a special seminar to be led by the renowned professor. A great number of young mathematicians were introduced to the art of teaching by "old Schellbach" at these seminars. Many of these young men went on to become truly outstanding secondary school teachers.

[In mathematics, Schellbach is today chiefly remembered for his elegant analytical solution of Steiner's generalization of Malfatti's problem, wherein, given three conics on a conicoid, one is to determine three others which shall touch two of the given and two of the required conics.]

47° *Schellbach previsioned.*   Novalis (1772–1801) preceded Schellbach in proclaiming mathematics to be a religion, and mathematicians to be the blessed priests of the religion. He once commented: "The life of God is mathematics; all divine ambassadors must be mathematicians. Pure mathematics is religion. Mathematicians are the only blessed people."

On mathematics as a branch of theology, see F. De Sua's discerning remark at the conclusion of Item 294° of *In Mathematical Circles.*

For some further, and sometimes penetrating, remarks on mathematics made by Novalis, see Item 223° of *Mathematical Circles Revisited.*

48° *The danger of mathematics.* Prinz Kraft zu Hohlenlohe-Ingelfingen (1827–1892), a Prussian general of artillery and author of several books on military science, studied under Karl Schellbach. Observing Professor Schellbach, Prinz Kraft said, "Mathematics is indeed dangerous in that it absorbs students to such a degree that it dulls their senses to everything else."

49° *A homework assignment.* To punish a student for a minor infraction, Schellbach gave the student a homework assignment which on completion the lad was to bring to the professor's home. The student did this, but found the professor was not available. He therefore gave the assignment to one of the professor's daughters. The next day in class the following dialogue took place:

Schellbach: "Why didn't you appear?"
Student: "I was there, Professor, and delivered the work. Someone said you were indisposed."
Schellbach: "To whom did you give the work?"
Student: "To one of your daughters, Professor."
Schellbach: "Which one?"
Student: "I don't know her name, Professor."
Schellbach: "Was she pretty?"
Student: "Yes, Professor."
Schellbach: "Well, then, that was Florence. Very good."

50° *Beauty and truth.* Georg Wilhelm Friedrich Hegel (1770–1831), the eminent German philosopher, said, "Who does not know the works of the ancients dies without knowing *beauty.*" Karl Schellbach responded, "Who does not know the works of the mathematicians and scientists dies without knowing *truth.*"

51° *Hegel speaks out.* The discovery [of the small planet Ceres] was made by G. Piazzi of Palermo; and it was the more interesting as its announcement occurred simultaneously with a publication by Hegel in which he severely criticized astronomers for not paying more attention to philosophy, a science, said he, which would at once have shown them that there could not possibly be more than seven planets, and a study of which would therefore have prevented an absurd waste of time in looking for what in the nature of things could never be found.

—W. W. R. BALL
A Short History of Mathematics, p. 448.

52° *A man of wide interests.*   The mathematician Hermann Günther Grassmann of Stettin, Germany, had very broad intellectual interests. He was not only a teacher of mathematics, but of religion, physics, chemistry, German, Latin, history, and geography. He wrote on physics and composed school texts for the study of German, Latin, and mathematics. He was a copublisher of a political weekly in the stormy years of 1848 and 1849. He was interested in music and in the 1860s he was an opera critic for a daily newspaper. He prepared a philological treatise on German plants, edited a missionary paper, investigated phonetic laws, wrote a dictionary to the Rig-Veda and translated the Rig-Veda in verse, harmonized folk songs in three voices, composed his great treatise *Ausdehnungslehre,* and brought up nine of his eleven children. His father was a teacher of mathematics and physics in the gymnasium at Stettin. His son Hermann Grassmann (b. 1859) also became a mathematician. The father wrote two books on mathematics, and the son wrote a treatise on projective geometry.

It was at the age of fifty-three that, through general lack of appreciation of his remarkable work, *Ausdehnungslehre,* Grassmann gave up mathematics and directed the bulk of his intellectual energies to the study of Sanskrit, achieving in philology splendid results that were much better appreciated than his contributions to mathematics. Grassmann spent his entire life in his native city of Stettin, except for the years from 1834 to 1836, when he taught mathematics in an industrial school in Berlin, having succeeded Jacob Steiner to the post.

53° *A good tactic.*   A special feature of Grassmann's classes was the so-called *Plauderminuten* (literally, "minutes for idle chatter"). Whenever he and the conscientious students in the class had worked steadily and intensely for some time on mathematics, Grassmann would call a halt and give his students and himself a short pause for free conversation. Grassmann would bring this about simply by sitting on the rostrum and calling, "Plauderminute."

54° *A doubtful tactic.*   One day in class, when some troublemakers were being unruly and were disturbing the entire class, Grassmann's pleas and admonitions went unheeded. Instead of taking disciplinary action, the professor descended from the podium and began praying long and loud to God, asking that He not take the ill behavior of the students into His account of their sins. He prayed that the students should seek to better themselves through His Grace, and so on. It is unlikely that this action yielded more than a temporary solution to the disciplinary problem.

55° *Forgiveness and a kiss.*   It seems that one day a student of the Obertertia sorely vexed Grassmann during a course in religious thought. As a result Grassmann was much too angry to give his customary prayer at the close of the hour. The student, noting the omission, became remorseful, and after class went repentantly to Grassmann and asked to be forgiven. Naturally the kind teacher forgave the boy and kissed him.

SEVEN MATHEMATICIANS AND A POET

THE seven mathematicians János Bolyai, Peter Gustav Lejeune Dirichlet, James Joseph Sylvester, Arthur Cayley, Carl Gustav Joseph Jacobi, William Thomson (Lord Kelvin), Augustus De Morgan, and the poet Henry Wadsworth Longfellow were all born in the first quarter of the nineteenth century.

56° *A misjudgment.*   "You should detest it just as much as any evil practice; it can deprive you of all your leisure, your health, your rest, and the whole happiness of your life. This abysmal darkness might perhaps devour a thousand Newtons; it will never be light on earth . . . ."

[From a letter of 1820 written by Farkas Bolyai (1775–1856) to his son János Bolyai (1802–1860), attempting to discourage the latter from investigating the parallel postulate.]

57° *Dirichlet as a teacher.*   Peter Gustav Lejeune Dirichlet (1805–1859) has been described as possessing a noble, sincere, human, and modest disposition, but he seemed unable to communicate with young minds. When a schoolmate expressed envy because Dirichlet's son could always receive help from his gifted father, the son gave this lamentable but memorable reply: "Oh! My father doesn't know the little things anymore." Dirichlet's waggish nephew, Sebastian Hensel, wrote in his memoirs that the mathematics instruction he received in his sixth and seventh years at the gymnasium from his uncle was the most dreadful experience of his life.

—ELWOOD EDE
(freely translated from W. Ahrens, *Mathematiker-Anekdoten*)

58° *Dirichlet as a correspondent.*   Dirichlet was polite and kind but very reserved, and he seldom committed himself to written communication. Once, when a member of the family received a letter from him, the letter was kept as "an extremely rare document."

When Dirichlet's first child arrived, Dirichlet failed to write of the event to his father-in-law, who was living in London at the time. The father-in-law, when he finally found out, commented that he thought Dirichlet "should have at least been able to write $2 + 1 = 3$." This witty father-in-law was none other than Abraham Mendelssohn, a son of the philosopher Moses Mendelssohn, and father of the composer Felix Mendelssohn.

—ELWOOD EDE
(freely translated from W. Ahrens, *Mathematiker-Anekdoten*)

31

59° *A doctored notice.*   Professor G. Arendt attended, in the summer of 1854 at the University of Berlin, a series of lectures by Dirichlet which Arendt published fifty years later in 1904. The following anecdote in connection with those lectures comes from the professor's letters.

It seems that, during the semester, Dirichlet was forced to postpone the lecture on "Definite integrals" because he became ill. He informed his students by posting the following notice, written in his tiny script, on the door of Room 17:

> Because of illness I cannot lecture today
> Dirichlet

Dirichlet's absence continued for almost a week while the notice remained on the door. Finally one of the students doctored up the message so that it took the form of a definite integral as follows:

$$\int_{Easter}^{Michaelmas} (\text{Because of illness I cannot lecture today}) \; d(1 \text{ Frdor})$$

> Dirichlet

It must be realized that 1 Friedrichsd'or was the customary honorarium for a semester lecture, and that Michaelmas is the feast of St. Michael celebrated on September 29.

—ELWOOD EDE
(freely translated from W. Ahrens, *Mathematiker-Anekdoten*)
[One is reminded of the doctoring of a similar notice posted one day by Professor William Thomson (Lord Kelvin) on his lecture-room door. See Item 348° of *Mathematical Circles Revisited.*]

60° *Sylvester on graph theory.*   J. J. Sylvester (1814–1897) explained the abstract nature of graph theory as follows: "The theory of ramification is one of pure colligation, for it takes no account of magnitude or position; geometrical lines are used, but these have no more real bearing on the matter than those employed in genealogical tables have in explaining the laws of procreation."—LEO MOSER

61° *Absent-mindedness.* The following anecdote, although clearly not quite factual, illustrates the reputation for absent-mindedness that Sylvester enjoyed.

Upon arrival from England at Johns Hopkins University, Sylvester was asked to give a major address at the university. He had prepared a manuscript for the address, but after he was introduced and called to the podium he could not locate his manuscript. He asked that the lecture be postponed. On return to his office he made a thorough search but still could not find his manuscript. He decided that he must have left it in England, so he returned overseas to look for it. When he arrived at his home he again made a diligent search for the lost manuscript and at long last he found it—in his pocket.—LEO MOSER

62° *Acquiring a new surname.* James Joseph, the youngest of a number of children born to Abraham Joseph, was born in London on September 3, 1814. The eldest son eventually migrated to the United States, where, for some reason not now known, he took on the surname Sylvester. The rest of the family soon followed the example, and ever after James Joseph was known as James Joseph Sylvester or, more briefly, as J. J. Sylvester.

63° *Sylvester's first contact with the United States.* Sylvester's American brother, who was an actuary, suggested to the Directors of the Lotteries Contractors of the United States that they submit a difficult problem in arrangements that was bothering them to his younger brother James Joseph, then only sixteen years old. James's complete and satisfying solution of the problem caused the Directors to award the young mathematician a prize of five hundred dollars.

64° *Sylvester and music.* In addition to a deep interest in poetry, Sylvester was also interested in music and was, in fact, an accomplished amateur. At one time he took singing lessons from the famous French composer Charles François Gounod and on occasions entertained at workingmen's gatherings with his songs.

It is said that he was prouder of his high C in singing than he was of his invariant theory in mathematics.

65° *Unconscious plagiarism.*   Plagiarism in mathematics is not always deliberate but sometimes occurs quite unconsciously. Consider the following cases, cited by J. J. Sylvester of himself, in *The Philosophical Transactions of the Royal Society* (1864), p. 642.

"Professor Cayley has since informed me that the theorem about whose origin I was in doubt, will be found in Schläfli's *De Eliminatione.* This is not the first unconscious plagiarism I have been guilty of towards this eminent man whose friendship I am proud to claim. A more glaring case occurs in a note by me in the *Comptes Rendus,* on the twenty-seven straight lines of cubic surfaces, where I believe I have followed (like one walking in his sleep), down to the very nomenclature and notation, the substance of a portion of a paper inserted by Schläfli in the *Mathematical Journal,* which bears my name as one of the editors upon the face."

66° *The only instance on record when Cayley lost his temper.* Arthur Cayley (1821–1895) was a singularly serene, even-tempered, and always unruffled individual. Only once is it known that he lost his calmness. It seems that one day when he and his friend Sylvester were discussing the theory of invariants in Cayley's law rooms, an office boy came in and handed Cayley a sheaf of legal papers for perusal. A glance at this pile of dismal work coming in the middle of his interesting mathematical discussion with Sylvester caused Cayley to flare up. He seized the batch of papers from the interrupting boy and, with an exclamation of disgust, hurled the material on the floor, and then went on talking mathematics with Sylvester.

67° *Cayley's style.*   Shortly after Cayley died in 1895, A. R. Forsyth wrote an "appreciation" of the great mathematician and his work in *The Proceedings of the London Royal Society,* Vol. 58 (1895). In this paper Forsyth, writing of Cayley's style, says:

"When Cayley had reached his most advanced generalizations he proceeded to establish them directly by some method or other, though he seldom gave the clue by which they had first been obtained: a proceeding which does not tend to make his papers easy reading. . . .

"His literary style is direct, simple and clear. His legal training had an influence, not merely upon his mode of arrangement but also upon his expression; the result is that his papers are severe and present a curious contrast to the luxuriant enthusiasm which pervades so many of Sylvester's papers."

68° *More on the Jacobi brothers.*   Item 344° of *In Mathematical Circles* reports the confusion that existed between the mathematician Carl Gustav Joseph Jacobi (1804–1851) and his brother Moritz Hermann Jacobi (1801–1874), the inventor of electrotyping. Once, when mistakenly taken for his brother, the mathematician Carl replied, "Pardon me, sir, I am not me. I am my brother."

The confusion between the Jacobi brothers was well known to many of Carl's friends, who, aware of Carl's penchant for droll humor, often jokingly took advantage of the mix-up. For example, Carl greatly admired the English astronomer and mathematician Mary Sommerville (1780–1872), and, when both were in Rome, he frequently called on her. Each time he did so, she would address him as, "Monsieur, votre frère [Sir, your brother]," which always caused Carl to chuckle.

Carl was a close friend of Professor and Mrs. Dirichlet. Once the latter gave Jacobi a Christmas package on which she had written, "al fratello del celebro Jacobi [to the brother of the celebrated Jacobi]."

On another occasion, when a very proper lady asked Carl whether he was a brother of the famous Jacobi, he replied, "Gracious, no, lady, he is my brother." Here one is reminded of the similar story, reported in Item 163° of *Mathematical Circles Squared*, about Henri Poincaré and his famous cousin Raymond Poincaré.

69° *Jacobi's drollery.*   Jacobi's drollery broke out on numerous occasions. Thus there was the time when he and a number of scholars attended a lecture given by Jacobi's friend, the astronomer F. W. Bessel (1784–1846). At the conclusion of the lecture, some of the scholars complained that Bessel had not really said anything new. Jacobi replied, "Yes, my friends, if Bessel had not explained his lecture to me during an hour-long walk this morning, I would not have understood it either."

—ELWOOD EDE
(freely translated from W. Ahrens, *Mathematiker-Anekdoten*)

70° *The Hävernick affair.*   Another instance of Jacobi's droll wit occurred when he was teaching at the University of Königsberg. At that time, the orthodox theologian Hävernick received an appointment to the "Albertina." Now the students disliked Hävernick because a decade earlier, while a student at Halle, he had served as an informant in the persecution of the rationalistic theologians Gesenius and Wegscheider. The result was that the students provoked an incident when Hävernick gave his first lecture at Königsberg. By plan, the entire student body crowded into the lecture hall, giving Hävernick a false feeling of great popularity, but as soon as Hävernick uttered his first syllable, the students stormed speedily and noisily out of the hall, leaving the deflated lecturer alone at his podium. The incensed Hävernick caused a hearing of the affair to be held by the University Senate, with an aim toward dismissal of the rude students. The prorector of the University, the anatomist Burdoch, concocted an apology to Hävernick, claiming that it was the steam and intense heat of the room that caused the students to rush out of the hall. At this point, Jacobi commented that, anyway, since only three students are needed to comprise a college, only the last three to leave the hall should be punished for dissolving the college.

—ELWOOD EDE
(freely translated from W. Ahrens, *Mathematiker-Anekdoten*)

71° *A Jacobi tactic.*   Jacobi had little respect for arrogant or pompous people. He once claimed, "If I have to speak with a tyrant or a foreign minister, I take pains to disappoint him beforehand."

72° *A gifted student.*   Jacobi was a brilliant student and won the admiration of all his teachers. The wide range of his talents is illustrated by the fact that in those early years he composed poetry. He once recited one of his own poems at a school assembly; the poem was written in Greek.

73° *Advice to a student.* S. P. Thomson, in his *Life of Lord Kelvin* (1910), tells the following amusing story about the assistant to the famous mathematical-physicist Sir William Thomson (Lord Kelvin) (1824–1907).

The father of a new student, when bringing his son to the University, drew Professor Thomson's faithful assistant, Donald McFarlane, to one side and besought the assistant to tell him what his son must do to stand well with the professor. "You want your son to stand weel with the Profeessorr?" asked McFarlane. "Yes," replied the father. "Weel, then, he must have a guid bellyful o' mathematics!"

74° *Obvious, but for a different reason.* Sir William Thomson (Lord Kelvin) often, in his writings, prefaced· some mathematical statement with the remark "it is obvious that" to the perplexity of his mathematical readers, for the statement was certainly not obvious from the mathematics that preceded it on the page. To Thomson the statement was obvious for *physical reasons,* and these reasons usually did not suggest themselves to the mathematicians, however competent they might be.

75° *A law of life?* In his charming *A Budget of Paradoxes,* Augustus De Morgan (1806–1871) narrates the following whimsical anecdote.

"A few days afterwards, I went to him [the actuary referred to in Item 306° of *In Mathematical Circles*] and very gravely told him that I had discovered the law of human mortality in the Carlisle Table, of which he thought very highly. I told him that the law was involved in this circumstance. Take the table of the expectation of life, choose any age, take its expectation and make the nearest integer a new age, do the same with that, and so on; begin at what age you like, you are sure to end at the place where the age past is equal, or most nearly equal, to the expectation to come. 'You don't mean that this always happens?'—'Try it.' He did try, again and again; and found it as I said. 'This is, indeed, a curious thing; this *is* a discovery!' I might have sent him about trumpeting the law of life: but I contented myself with informing him that the same thing would happen with any table whatsoever in which the first column goes up and the second goes down; . . .''

76° *De Morgan anagrams.*   De Morgan relates, in his *A Budget of Paradoxes,* that someone devised 800 anagrams on his name, of which he had seen about 650. Most of the anagrams are in Latin; here are two that are in English:

$$\text{Great gun! do us a sum!}$$
$$\text{Go! great sum! } \int a^u{}^n \, du$$

The first he regarded as a sneer at his pursuit and the second as more dignified.

77° *A problem in projectiles.*   There have been instances wherein a good mathematical problem has been suggested by some passage in a piece of literature. Perhaps the most famous such instance, historically, is the problem of the duplication of a cube, which seems to have been first suggested by a passage written by an unschooled and obscure Greek poet telling the story of the mythical King Minos and the cubical tomb erected for his son Glaucus. And one recalls how Euler was led to a problem in mechanics by a line in Virgil's *Aeneid.* These instances are told in Item 246° of *In Mathematical Circles,* along with that of an intriguing geometry problem that appeared some years ago in the Problem Department of *The American Mathematical Monthly,* suggested by some lines in Jan Struther's *Mrs. Miniver.* There are so many such instances that one could write an interesting little monograph on problems arising from literature. Here is another one, found among William Walton's *Collection of Mechanical Problems* and stemming from *The Song of Hiawatha* by Henry Wadsworth Longfellow (1807–1882). This popular poem was published in 1855. Based on Henry Rowe Schoolcraft's two informative books on the Indian tribes of North America and written in the catching trochaic metrics of the Finnish epic *Kalevala,* the poem was an immediate success. Here are the lines of the poem that caught William Walton's attention back in the mid-nineteenth century.

> Swift of foot was Hiawatha;
> He could shoot an arrow from him,
> And run forward with such fleetness,
> That the arrow fell behind him!
> Strong of arm was Hiawatha;
> He could shoot ten arrows upward,
> Shoot them with such strength and swiftness,
> That the tenth had left the bow-string
> Ere the first to earth had fallen.

From the above poetic lines, Walton proposed the following problem:

Suppose Hiawatha to have been able to shoot an arrow every second, and, when not shooting vertically, to have aimed so that the flight of the arrow might have the longest range. Prove that it would have been safe to bet long odds on him if he entered the Derby.

CHARLES HERMITE

THE two fundamental mathematical results due to Charles Hermite that are of most popular interest are his solution in 1858 of the general quintic equation by means of elliptic functions, and his proof in 1873 of the transcendence of the number e. Hermite's success with the quintic equation later led to the fact that a root of the general equation of degree n can be represented in terms of the coefficients by means of Fuchsian functions, and the method he employed to prove that e is transcendental was employed by Lindemann in 1882 to prove that π also is transcendental.

Charles Hermite was born in Dienze on December 24, 1822, and he died in Paris on January 14, 1901. Though not a prolific writer, most of his papers deal with questions of great importance and his methods exhibit high originality and wide usefulness. His collected works, edited by Émile Picard, occupy four volumes. Some facts about him of a more personal nature can be learned from the following seven anecdotes, which may make up for the grave deficiency of his almost complete absence in our previous three trips around the mathematical circle.

78° *Hermite's infirmity.*   Mankind has always been plagued by infirmities, and the mathematicians have not been exempt. Recall that Saunderson was blind almost from birth, that Euler spent the latter part of his life in blindness, that Clifford and Abel died early of tuberculosis, that Hilbert suffered from pernicious anemia. The list can be extended.

Charles Hermite (1822–1901) was born with a deformity of his right leg and was accordingly lame all his life, requiring a cane to get about. An infirmity can easily mar one's disposition. Such, however, was not the case with Hermite, who uniformly maintained the sweetest of dispositions, causing him to be loved by all who knew him. One good result of Hermite's deformity was that it very successfully barred him from any kind of a military career. One bad result was that after one year at the École Polytechnique he was dropped from further study because the authorities claimed that his lame leg rendered him unfit for any of the positions open to successful students of the school.

79° *Hermite's kindness to those climbing the ladder.*   A number of eminent mathematicians have exhibited great generosity to younger men struggling for recognition. Charles Hermite is regarded as unquestionably the finest character of this sort in the entire history of mathematics.

80° *The yardstick of examinations.*   Hermite and Galois, both misfit alumni of the Louis-le-Grand lycée, exhibited a great indifference to the elementary mathematics of the classroom, and accordingly, consistently did poorly on all their school examinations. Professor Richard of the school had tried to save Galois, but failed; he also came to the rescue of Hermite, and this time fortunately succeeded. Anyone who detests examinations will feel for Hermite and Galois, and the careers of these two outstanding mathematicians may create some doubts in the minds of those who advocate examinations as a reliable yardstick for measuring one's intellectual merit.

While at Louis-le-Grand, Hermite had two papers, one of quite exceptional quality, accepted by the *Nouvelles annales de mathématiques,* a journal founded in 1842 and devoted to the interests of students in the higher schools. Professor Richard felt compelled to confide to Hermite's father that Charles was "a young Lagrange."

Later in 1842, while still twenty years of age, Hermite took the entrance examinations for the École Polytechnique, and although he was unquestionably the mathematical superior of some of his examiners, he had the humiliating experience of coming in sixty-eighth in order of merit, just barely meeting the entrance requirements. Ironically enough, Hermite's first academic post was an appointment, in 1848, as an examiner for admissions at the École Polytechnique, the very school that five years earlier had almost failed to admit him.

81° *A slight variation.* Many young mathematicians have married daughters of their professors. The reader may recall, from Item 322° of *Mathematical Circles Squared,* the resulting peculiar law of genetics: "Mathematical ability is not inherited from father to son, but from father-in-law to son-in-law." Hermite varied the usual procedure somewhat; he married, not a daughter, but the sister of one of his professors—in 1848 Hermite married Professor Bertrand's sister Louise.

82° *Hermite's conversion.* In the early part of his life Hermite, like many French scientists of his time, was a tolerant agnostic. But in 1856 he suddenly fell dangerously ill. While in his debilitated condition, Cauchy, who had deplored the young man's open-mindedness in religion, worked on the weakened Hermite and converted him to Roman Catholicism. Hermite lived the rest of his life a devout and contented Catholic.

41

83° *Hermite's mysticism.*   The whole question of mathematical existence is a highly controversial one. For example, do mathematical entities and their properties already exist in a sort of timeless twilight land of their own, and we, wandering about in that land, accidentally discover them? In this twilight land the medians of a triangle are, and always have been, concurrent in a point trisecting each median, and someone, probably in ancient times, wandering about in his mind in the twilight land, came upon this already existing property of the medians of a triangle. In the twilight land, many other remarkable properties of geometrical figures have always existed, but no one has yet stumbled upon them, and may not for years, if ever. In the twilight land, the natural numbers and their host of pretty properties already exist, and always have, but these properties will become existent in the real land of man only when someone wandering about in the twilight land comes upon them.

Pythagoras entertained the above idea of mathematical existence, as have many mathematicians after him. Hermite was a confirmed believer in the twilight land of mathematical existence. To him, numbers and all their beautiful properties have always had an existence of their own, and occasionally some mathematical Columbus stumbles upon one of these already existing properties and then announces his *discovery* to the world.

For more on discovery versus invention in mathematics, the reader may consult Item 355° of *Mathematical Circles Squared.*

84° *Hermite's internationalism.*   In this world of internecine strife and nationalistic jealousies, too many scientists look upon the scientific accomplishments of an enemy land as *bad* science, while that of their own land is *good* science. This narrow viewpoint has not bypassed the mathematicians. Thus, during World War II, many mathematicians of the allied nations looked upon "German" mathematics as somehow deformed and inferior, and many of the German mathematicians looked upon "Jewish" mathematics as somehow pernicious and evil. Hermite could never understand such politically partisan views of mathematics and science, and though he was a strong French patriot, to him mathematics and science were just mathematics and science, with no national or religious confines.

85° *On Cayley.*   Writing in the *Comptes Rendus,* t. 120 (1895), p. 234, shortly after Cayley's death, Hermite said: "The mathematical talent of Cayley was characterized by clearness and extreme elegance of analytical form; it was reinforced by an incomparable capacity for work which has caused the distinguished scholar to be compared with Cauchy."

LEWIS CARROLL

ALL   mathematicians admire Lewis Carroll (Charles Lutwidge Dodgson). Here are a few more stories about him.

86° *Lewis Carroll and the kitten.*   Lewis Carroll (1832–1898) had a great empathy for animals. One time, when away from home, he came upon a kitten with a fishhook caught in its mouth. He carried the kitten to a doctor's house, where he held and comforted the animal while the doctor removed the hook by first snipping off the barbed end. Upon learning that the kitten did not belong to Carroll, the doctor declined payment. Lewis Carroll then carried the kitten back to the street where he had found it.

87° *Lewis Carroll and the horses.*   On another occasion, Lewis Carroll came upon some horses that were being worked with checkreins on them. The discomfort to the horses caused by the checkreins was evident and instantly roused Carroll's compassion. He spoke to the man working the horses, and so convincingly put the case against the use of checkreins that the man removed them. The horses, allowed the natural use of their necks, performed their work much more efficiently and quickly.

88° *Lewis Carroll and meat.*   In Lewis Carroll's day, the butchering of animals for meat was done with little regard for the animals' physical feelings, and Carroll urged that methods of painless death for animals be adopted in the butchering process.

89° *An incongruous remedy.*   Professor York Powell, Regius Professor of Modern History at Oxford, has recorded a comic story told to him by Professor Dodgson (Lewis Carroll). A small child on being put to bed called to its nurse, "Nursey, my feet, my feet." The nurse took the child out of its cot and bathed his legs with vinegar and hot water, and gave him some warm milk. Upon putting the child back to bed, it once again cried out, "Nursey, my feet, my feet." So the nurse once again removed the child from his cot, but could find nothing amiss. To be on the safe side, however, she again bathed the child's legs with vinegar and hot water and dried them very carefully, and then put the now very sleepy youngster back into bed again. Immediately came the cry, "Nursey, my feet, my feet." So she took a light to examine the cot and discovered that the child's older brothers had, as a prank, short-sheeted the bed so that the child could not comfortably straighten out his legs.

90° *Lewis Carroll's fireplace.*   There was a fireplace in Lewis Carroll's study at Christ Church. Around the two sides and the top of the fireplace were sixteen square tiles and, across the middle of the top, a long rectangular tile. The square tiles were designed with pictures of animals, and the long tile above the fireplace-opening contained a ship. When young people visited Carroll in his study he would make the animals on the tiles carry on long and amusing conversations among themselves. He explained that the bird with its beak running through a fish, the dragon hissing defiance over its left shoulder, and others, represented the various ways in which he was accustomed to receive his guests.

The ship in the center above the fireplace became the famous vessel that the Bellman steered, often with difficulty, since "the bowsprit got mixed with the rudder sometimes." But

"The principal failing occurred in the sailing,
And the Bellman, perplexed and distressed,
Said he *had* hoped, at least, when the wind blew due East,
That the ship would *not* travel due West!"

The animals in the square tiles came to play a part in a number of Carroll's works. Thus in the bottom right corner was the Beaver, the only animal that the Butcher in *The Hunting of the Snark* knew how to kill. In the top right corner was the Eaglet, one of the competitors in the Caucus Race in *Alice in Wonderland,* and below it was the Gryphon. In the two uppermost tiles on the left side were the Lory and the Dodo, also of Caucus Race fame. The bottom left tile showed the Fawn that couldn't remember its name in *Alice Through the Looking Glass.*

45

QUADRANT TWO

*From the antithesis of Plato
to Artin on graphs*

A MELANGE

WE enter the second quadrant with stories concerning an assortment of men—from Francis Bacon to G. H. Hardy.

91° *The antithesis of Plato.* Thomas Macaulay, in his essay of 1837 on Francis Bacon (1561–1626), says: "Assuming the well-being of the human race to be the end of knowledge, he [Lord Bacon] pronounced that mathematical science could claim no higher rank than that of an appendage or an auxiliary to other sciences. Mathematical science, he says, is the handmaid of natural philosophy; she ought to demean herself as such; and he declares that he cannot conceive by what ill chance it has happened that she presumes to claim precedence over her mistress." And then, a bit further on, Macaulay supports Bacon by saying: "If Bacon erred here [in valuing mathematics only for its uses], we must acknowledge that we greatly prefer his error to the opposite error of Plato. We have no patience with a philosophy which, like those Roman matrons who swallowed abortives in order to preserve their shapes, takes pains to be barren for fear of being homely."

In truth, both Bacon and Macaulay were quite ignorant of the real nature and accomplishments of mathematics. As Augustus De Morgan said in his *A Budget of Paradoxes* of 1872, "If Newton had taken Bacon for his master, not he, but somebody else, would have been Newton."

92° *Hard to believe.* Most of our knowledge of Samuel Johnson (1709–1784), the great critic, man of letters, and lexicographer, has come to us from the remarkable biography of him written by his friend, the lawyer James Boswell. In this biography Boswell says, "When Dr. Johnson felt, or fancied he felt, his fancy disordered, his constant recurrence was to the study of arithmetic."

93° *A contrast of styles.* W. W. Rouse Ball, in his engaging *A Short Account of the History of Mathematics,* interestingly contrasts the styles of Lagrange, Laplace, and Gauss. "The great masters of

49

modern analysis are Lagrange, Laplace, and Gauss, who were contemporaries. It is interesting to note the marked contrast in their styles. Lagrange is perfect both in form and matter, he is careful to explain his procedure, and though his arguments are general they are easy to follow. Laplace on the other hand explains nothing, is indifferent to style, and, if satisfied that his results are correct, is content to leave them either with no proof or with a faulty one. Gauss is as exact and elegant as Lagrange, but even more difficult to follow than Laplace, for he removes every trace of the analysis by which he reached his results, and studies to give a proof which while rigorous shall be as concise and synthetical as possible.''

94° *Overshadowed.*   There have been occasions when an eminent mathematician was, during his lifetime, overshadowed by a much less gifted relative. Recall Henri Poincaré and his cousin Raymond Poincaré (see Item 162° of *Mathematical Circles Squared*) and C. G. J. Jacobi and his brother M. H. Jacobi (see Item 344° of *In Mathematical Circles* and Item 68° of the present volume). Another such instance is that of Augustus Ferdinand Möbius and his son. The father was born in Saxony in 1790 and, after an initial misstart preparing for the law, turned to mathematics. As a student of Gauss in 1813–14, he received training in astronomical observation and calculation. In 1815 he became a lecturer at the University of Leipzig, where he remained the rest of his life. In 1816 he became Extraordinary Professor of Astronomy and then, later, the Director of the Pleissenburg Observatory. In 1844 he was named Professor of Higher Mathematics and Astronomy at the university. He died, after a distinguished career, in 1868.

According to Heinrich Tietze, Möbius had a son who became a well-known neurologist, and whose book on the "physiologically weaker mind of women" won much more notice than did the sounder mathematical works of his father.

95° *A bright pupil.*   Even as a student at the Johanneum in Lüneberg, Riemann had so distinguished himself in mathematics that his mathematics instructor, Schmalfuss, allowed the boy to occupy himself in any manner he chose during the mathematics hour. Schmalfuss, an able mathematician in his own right, realized that Reimann had already mastered everything in mathematics that the school could offer.

96° *A perfectionist.* Riemann greatly excelled in all of his subjects at the gymnasium, and he was indisputably very diligent, but, because he was a perfectionist, he rarely handed in any of his important assignments on time. His dormitory resident, a stern master, stood over the boy to be sure he didn't loaf. Still the assignments were late. Constantly dissatisfied with the quality of his work, Riemann would tear up one paper after another, and proceed anew. As a consequence, punishment was regularly meted out to the youngster, and he was often sent to the school prison cell to complete a late assignment. While thus incarcerated, he would ultimately produce a near-perfect paper, and he consistently earned top grades.

Riemann exhibited his trait of slow and patient striving for perfection in later years. He took an unusually long time to complete his doctorate to his own satisfaction, with the result that it was also a long time before he was promoted to a full professorship at Göttingen University.

97° *A beautiful result.* Lazarus Fuchs (1833–1902), of differential equations fame, was born near Posen. He began his long teaching career at the University of Berlin, then taught at the Universities of Greifswald, Göttingen, and Heidelberg, and returned to Berlin as a full professor in 1884. It was in 1865 that he initiated a new theory of linear differential equations.

There is a frequently told story about a lecture given by Fuchs in his course on differential equations. It seems that he often appeared in class with his lecture unprepared, preferring to work out details at the blackboard. The result was that he occasionally made errors or cornered himself. During one of his lectures, while trying to derive a certain relation that he needed, he filled the board with a long series of complicated equations obtained by a sequence of involved substitutions. All at once the mass of equations simplified, and Fuchs suddenly arrived at the identity $0 = 0$. Puzzled and embarrassed, the professor paced a bit before the blackboard, muttering that there must be something wrong somewhere. Finally he turned to the class, winked, and said, "But no, zero equals zero is really a very beautiful result." Here the lecture terminated. At the following class meeting, Fuchs circumvented the whole problem by simply writing down the result he had originally wanted, and proceeded from there.

98° *A versatile genius.* Hermann Ludwig Ferdinand von Helmholtz was born in Potsdam in 1821. He started his career as a student of medicine in Berlin and was appointed assistant surgeon in the Charité Hospital there in 1842. The following year he went to Potsdam as a military surgeon, but returned to Berlin in 1848 to teach anatomy at the Academy of Art. In 1849 he was called to the chair of physiology at Königsberg. Six years later he moved to Bonn as professor of anatomy and physiology. In 1858 he was appointed professor of physiology at Heidelberg. In physiology he did pioneering work on the human nervous system and was the first to measure the speed of nerve impulses. Among other things, he invented the ophthalmoscope used by doctors to examine eyes. His research work in physiology demanded so much physics that he applied himself to that subject and became an outstanding physicist. In physics he conducted pioneering research in acoustics and in optics, helped to prove the law of conservation of energy, and developed electromagnetic theory. In 1871 he returned to Berlin as professor of physics, and in 1888 he became president of the Imperial Physics-Technical Institute in Charlottenburg. His creative work in physics made such demand on mathematics that he became a high-ranking mathematician of his time. In addition to work in applied mathematics, Helmholtz contributed to the field of geometry, particularly to non-Euclidean geometry. He died in Charlottenburg in 1894.

Of Helmholtz, William Kingdon Clifford wrote: "Helmholtz— the physiologist who learned physics for the sake of his physiology, and mathematics for the sake of his physics, and is now in the first rank of all three."

99° *Groping.* Alfred North Whitehead (1861–1947) once cautioned a student about a theory of logic. "You must take it with a grain of, er . . . um . . . ah . . ." For almost a minute Whitehead groped for the word, until the student suggested, "Salt, Professor?" "Ah, yes," Whitehead beamed, "I knew it was some chemical."—LEO MOSER

100° *Scepticism.* It has been claimed that only three people read the three volumes of Russell and Whitehead's monumental *Principia mathematica* in its entirety, namely Russell, Whitehead, and the proofreader. There has been some scepticism, however, about Russell and Whitehead belonging to the list.

—LEO MOSER

101° *Hilbert's memory.* Helmut Hasse, when on a visit to the University of Maine in March, 1972, told an amusing story about himself and David Hilbert (1862–1943). According to the story, Hasse once expressed to Mrs. Hilbert a desire to speak personally with the great mathematician. Mrs. Hilbert accordingly invited Hasse to tea one afternoon and then left him in the garden with her husband. Hasse soon launched into a discussion of class-field theory, a subject that had been created by Hilbert and that was of great interest to Hasse at the time. Hasse had written a report in the theory that had continued the earlier work done by Hilbert, and Hasse started to tell Hilbert of his added contributions. But Hilbert repeatedly interrupted Hasse, insisting that Hasse first explain the basic concepts and foundations of class-field theory. Hasse did so, and Hilbert grew enthusiastic and finally exclaimed, "All this is extremely beautiful; who created it?" And Hasse had to tell the astonished Hilbert that it was Hilbert himself who had created the beautiful theory.

One is reminded of the similar story told of J. J. Sylvester (see Item 353° of *In Mathematical Circles*).

102° *Hilbert and his boiled eggs.* During World War I, Hilbert was met on his way to the University by a colleague. Asked by the colleague, "How are you, Professor Hilbert?" he replied, "Oh,

53

rather bad." And when further asked, "What is the matter with you?" he said, "My wife is trying to murder me." The colleague was astonished to hear this. He said he could not believe it, because he knew how anxious Mrs. Hilbert was about her husband's welfare. But Hilbert told him that since their marriage his wife had boiled him *two* eggs each morning for breakfast, and the colleague would certainly understand how shocked he was this morning to get only *one* egg. (It was at the time when food rationing had set in in Germany.)—HELMUT HASSE

103° *Hilbert and his torn trousers.*   Hilbert used to go on a bicycle excursion, on a Saturday afternoon, with the students of his Seminar during the strawberry season. Now it was about this time of the year that his students had seen the Professor day by day coming to the University with a large rent in the seat of his trousers. The great respect they had for the Professor prevented them from drawing his attention to the tear. So they decided to wait until the impending excursion to the nearby village of Nikolausberg, where they were to revel in fresh strawberries and cream, and they then would say to him, as he descended from his bicycle, "Herr Professor, you have just now, in getting off your bicycle, torn your trousers." They did as planned, but Hilbert immediately answered, "But no, gentlemen, I have had this already for a fortnight."—HELMUT HASSE

104° *An unusual visit.*   Jacques Hadamard (1865–1963) has told that, as editor of a mathematics journal, he once received rather good papers from someone unknown to him, so he invited the person to dinner. The correspondent wrote that owing to circumstances beyond his control he could not accept the invitation, but he invited Hadamard to visit him. Hadamard did so and found to his great surprise that the author was confined to a criminal lunatic asylum. Apparently he was quite sane except for the murder of his aunts. His name was A. Bloch, and he was a very good mathematician.—L. J. MORDELL

105° *A conscientious reading.* Edmund Landau (1877–1938) was once sent a thesis to referee. He did not return it for a long time and several months later the sender met Landau and asked him if he had read the work. Landau replied, "Ya, Ich habe es grundlich durchgebletert [Yes, I have thoroughly paged through it]."—Leo Moser

106° *Place cards.* Landau gave an annual banquet for his graduate students. Instead of place cards he made up cards containing some characteristic phrase or formula from the subject of the student's thesis. One student had made no progress whatever towards a thesis topic and he wondered what would appear on his card. It turned out that he had no difficulty in finding his place. One card was quite blank.—Leo Moser

107° *Continuity.* Professor G. H. Hardy (1877–1947) did not believe in wasting time reviewing his earlier lectures—the students were expected to retain in their minds the material already covered. It chanced that a series of mathematical lectures given by Professor Hardy at Cambridge University was interrupted by the long vacation. On the first class meeting after the vacation, Professor Hardy advanced to the blackboard with chalk in hand and said, "It thus follows that. . . ."—Stanley B. Jackson

108° *Hundred percent achievers.* When I visited Hardy in his rooms in Trinity College in 1933, I saw three pictures on his mantelpiece, namely, one of Lenin, one of Jesus Christ, and one of Hobbs (the then star cricketer in England). Asked why there were exactly those three, Hardy said they were, in his opinion, the only important personalities who had achieved a hundred percent of what they wanted to achieve.—Helmut Hasse

109° *Hardy tries to outsmart God.* [In Item 298° of *Mathematical Circles Revisited,* we told George Polya's version of a story of how Hardy once attempted to outsmart God. Good stories often appear with variations, and this is true of the present story. Here

55

is Leo Moser's version of the story; it replaces the Riemann hypothesis of the former telling by Fermat's last theorem.]

Hardy claimed that he believed God existed but that God was his personal enemy. Once Hardy had to travel by air to France to attend a meeting. He feared air travel and so was very reluctant to go. Finally he did so, but he left a note on his desk to be opened by his secretary after his departure. When the note was opened, it was found to read: "I have just discovered a proof of Fermat's last theorem and will supply the details upon my return from France." His colleagues were quite excited about this and when Hardy came back they immediately pounced upon him. "What about the proof?" they asked. "Oh, that!" Hardy replied, "That was just insurance. I felt sure that God would spare me on this trip in order that no undue credit would go to me for having solved Fermat's last theorem."—LEO MOSER

ALBERT EINSTEIN

PROBABLY no mathematician of modern times has been more universally known and admired than Albert Einstein (1879–1955), and stories about this great man are always welcomed. Here are some more to add to those we have already told in our earlier trips around the mathematical circle.

110° *Home-town honors.*   Albert Einstein was born in Ulm, Germany, in 1879. Though the Einstein family moved from Ulm to Munich when Albert was only about a year old, one nevertheless wonders if the city of birth of the famous mathematical physicist has in any way honored its most outstanding citizen. For example, one might expect that the house in which Einstein was born has been preserved as an historical shrine, but that house was blasted to rubble in World War II and thus no longer stands. However, a street in Ulm was named Einsteinstrasse in honor of Einstein. But the Nazis could not bear to see a Jew honored in this way, and the new Nazi mayor of Ulm on his first day in office changed the name of the street to Fichtestrasse, after the eighteenth-century German philosopher, orator, and patriot Johann Gottlieb Fichte (1762–1814). Succeeding the defeat of the Nazis, the name Einstein-strasse was restored, and so it stands today.

111° *No sales resistance.*   There is a story about Einstein that says one day a large truck appeared before his modest two-story frame house in Princeton, New Jersey, containing a sizable elevator to be installed in the house. It seems that Einstein, who had absolutely no need or desire for such an expensive luxury, could never be rude to a salesman, and so some weeks earlier had allowed himself to be seduced into ordering the thing. Mrs. Einstein refused acceptance of the elevator and sent the loaded truck away. —Robert T. Hall

112° *The great stimulator.*   An enormous emotion was aroused in Albert Einstein very early in his twelfth year when he came into possession of a small textbook on Euclidean geometry. The book utterly absorbed his interests, and later, in his *Autobiographical Notes,* he wrote with rapture of "the holy little geometry book." He says: "Here were assertions, as for example the intersection of the three altitudes of the triangle in one point, which—though by no means evident—could nevertheless be proved with such certainty that any doubt appeared to be out of the question. This lucidity and certainty made an indescribable impression on me." We have here another instance of an eminent mathematician whose first great stimulus for mathematics was received from the enchanting material of Euclid's geometry.

113° *Mystery.*   One night in New York, scientist Albert Einstein attended a banquet given in his honor. Mrs. Einstein, ailing with a cold, did not accompany him.

It was a formal affair, with the men in white ties and the ladies in décolleté evening gowns.

When Einstein came home, he found his wife waiting up for him, eager to learn what had taken place. He began to tell her about the famous scientists who were present, but she cut him short.

"Never mind that," she said. "How were the ladies dressed?"

"I really don't know," replied Einstein. "Above the table, they had nothing on, and under the table I didn't dare look!"

—*Famous Fables* by EDGAR

114° *The weapons of World War IV.* Someone once asked Einstein what weapons will be used in World War III. "I don't know," he replied, "but I do know what weapons will be used in World War IV." "What will those weapons be?" he was asked. "Sticks and stones," he replied.

115° *A considerate man.* Dr. John Wheeler of Seattle, Washington, and one of America's leading theoretical physicists, taught at Princeton University during the early 1950s when Albert Einstein was also there. He recalls, "Dr. Einstein used to walk past my house on his way home from the university, and from time to time my children's cat would follow him home. As soon as he got there, he would phone to tell me that our cat was over at his house. He didn't want us to worry."

116° *A late bloomer.* Albert Einstein was unable—or perhaps unwilling—to speak until he was three years old. In school he had no facility at sums and had to be taught the multiplication table by raps on the knuckles. He was so mediocre as a pupil that at seven his teacher said "nothing good" would ever come of him. When he sought admission to the Polytechnic Institute in Zurich (now the Federal Institute of Technology), he failed to pass the entrance examination and had to prepare for a second attempt at entrance. After graduating from the Institute, he managed to secure a modest position at the patent office in Berne; his efforts to obtain the equivalent of a high school teaching post were consistently turned down.

Einstein's ineptness in simple arithmetic followed him through life, but it is believed that his late use of verbal communication led him to develop an extraordinary capacity for nonverbal conceptualization, so that the use of abstract concepts, rather than words, persisted into his adult life. Dr. Gerald Holton, professor of physics at Harvard University, believed Einstein's habit, from infancy on, of thinking in concepts rather than words played a key role in Einstein's scientific work. Once, when discussing the genesis of his ideas with a friend, Einstein commented: "These thoughts did not come in any verbal formulation. I rarely think in words at all. A thought comes, and I may try to express it in words afterward." Elsewhere he wrote: "The words or the language, as they are written or spoken, do not play any role in my mechanics of thought." The elements of Einstein's thought seem to have been sets of images that he could voluntarily reproduce or combine, or, sometimes, certain muscular reminders would be used instead of visual symbols.

117° *Two mementos.* There were two objects, Einstein claimed in later years, that played a special inspiration in his life. One of these was the little geometry book mentioned above in Item 112°; the other was a directional compass given to him by his father when he was four or five years old. The little book enthralled him as a boy because of the beautiful certainty of the deductive reasoning revealed in it; the compass mystified him as a youngster because of the all-pervading magnetic field that cleverly controlled the needle. Einstein was so impressed by these two objects that he kept them close to him all his life.

What has become of these two objects that so influenced Einstein's life? The little book is still in existence—in the filing cabinets at Princeton; the compass, on the other hand, has somehow vanished.

118° *Declining a presidency.* Einstein exerted long efforts on behalf of the creation of a Jewish national state. On May 15, 1948, the dream came true with the establishment of the independent State of Israel in Palestine. In 1952 the Israeli government asked Einstein to accept the presidency of that country as successor to Chaim Weizmann. Einstein sadly declined the honor, insisting that he was not fitted for such a position.

119° *Einstein, the sage.*   In his later years, Einstein became fixed in the public eye as something of a philosophical sage, and many people wrote to him for advice concerning their personal problems. One of the most poignant of these letters was one that he received in 1950 from an ordained rabbi. The rabbi wrote that he had tried in vain to comfort his nineteen-year-old daughter over the death of her sixteen-year-old sister. The surviving daughter found no consolation "based on traditional religious grounds," and so the concerned father thought that perhaps a scientist could help.

"A human being," wrote Einstein in his reply, "is a part of the whole, called by us 'Universe,' a part limited in time and space. He experiences himself, his thoughts and feelings as something separated from the rest—a kind of optical delusion of his consciousness. This delusion is a kind of prison for us, restricting us to our personal desires and to affection for a few persons nearest to us. Our task must be to free ourselves from this prison by widening our circle of compassion to embrace all living creatures and the whole of nature in its beauty. Nobody is able to achieve this completely, but the striving for such achievement is in itself a part of the liberation and a foundation for inner security."

120° *The deleted passage.*   Einstein was an impassioned humanitarian and internationalist and he regarded nationalism and patriotism as generators of evil aggressiveness.

On October 23, 1915, at the height of World War I and while living in Berlin, Einstein was invited by the Berlin Goethe Society to submit an essay for publication in the Society's journal. The ensuing correspondence brought Einstein face to face with German nationalism, and finally forced him to delete a passage from his article. In reply to the invitation Einstein wrote, "Of course, I will not be surprised, or even indignant, if you do not make use of my remarks. However, in that case, I ask you to send the same back to me."

The Goethe Society was indeed dismayed by Einstein's submitted essay, for it contained a passage in which Einstein equated patriotism with the worst of aggressive animal instincts. The outcome was that he was asked to delete this part from his article, for the concept of patriotism was the chief prop then upholding the morale of the German troops bogged down in muddy trenches along the various fronts of the war. Here is the passage that Einstein finally agreed to delete:

"One may ask oneself why it is that in peacetime, when the social system suppresses almost all expressions of virile pugnaciousness, the attributes and drives that during war generate mass murder do not disappear. In that respect it seems to me as follows:

"When I look into the home of a good, normal citizen I see a softly lighted room. In one corner stands a well-cared-for shrine, of which the man of the house is very proud and to which the attention of every visitor is drawn in a loud voice. On it, in large letters, the word 'Patriotism' is inscribed.

"However, opening this shrine is normally forbidden. Yes, even the man of the house knows hardly, or not at all, that this shrine holds the moral requisites of animal hatred and mass murder that, in case of war, he obediently takes out for his service.

"This shrine, dear reader, you will not find in my room, and I would rejoice if you came to the viewpoint that in that corner of your room a piano or a small bookcase would be more appropriate than such a piece of furniture which you find tolerable because, from your youth, you have become used to it."

Continuing, Einstein wrote: "It is beyond me to keep secret my international orientation," and he concluded by saying that the state, to which he happens to belong as a citizen, "does not play the least role in my spiritual life; I regard allegiance to a government as a business matter, somewhat like the relationship with a life insurance company."

Thus ended the offending passage that Einstein finally agreed to delete from his essay. He concluded the article, however, by saying: "But why so many words, when I can say everything in one sentence, and also a sentence that suits my being a Jew: Honor your master, Jesus Christ, not only with words and hymns, but above all thoroughly by your deeds."

121° *First brush with patriotism.* Einstein said he could never forget the open hatred his classmates in grade school had, under a misguided sense of neighborhood patriotism, for the students of a nearby school. Numerous unreasonable fist fights took place, many heads were battered in the frenzied battles, and bloodied "heroes" were idolized.

122° *Einstein as the hero of an opera.* The prominent East German composer Paul Dessau for almost twenty years carried around the idea of composing an opera built about the figure of Albert Einstein. The opera was not to be biographical, nor a setting in music of relativity theory, but a sort of morality play on a scientist's responsibility for destructive discoveries that he lets loose upon the world. Finally, in 1974, when eighty years old, Dessau saw his opera completed and performed. Under superb direction and with excellent singers, the opera *Einstein* enjoyed a very successful world premier at the East Berlin Staatsoper, conducted by Otmar Suitner of the San Francisco Opera Company.

The opera is unconventional, but contains fascinating music and a comic allegoric episode provided by *Hanswurst* (Germany's *Punch*) after each act. A single setting cleverly used with simple props and scenic units evokes the various atmospheres of the rapidly changed nineteen scenes.

The opera opens with an alarming prologue informing the audience that the Nazis (all appearing with their arms bloody to the elbows) are burning undesirable books on the square just outside the Staatsoper (where the burning actually took place), a site "you can visit during the intermission," and Scene One shows the infamous incident. Einstein's apartment is vandalized by storm troopers and the mathematical physicist leaves for the United States with his violin under his arm. One of Einstein's colleagues follows him, and then returns to Germany as an American officer to impress the services of a second mathematical physicist, a Nazi collaborator and organ virtuoso, to develop the atomic bomb. Act II ends when the bomb is dropped on Hiroshima. In Act III Einstein and his fellow scientists, guarded by MPs carrying machine guns (all United States authorities, including "The President"—presumably Roosevelt—appearing with their legs bloody to the knees) are haled before the United States Supreme Court to answer for their conduct in condemning the bomb, and Einstein is "sentenced" to everlasting fame. Einstein's former colleague leaves to join the Communist camp, and Einstein, now a bitter old man, destroys his latest theories before they can be put to inhuman use. It is in the allegoric *Hanswurst* intermezzi closing each act that the moral of the opera is provided.

L. J. MORDELL

THERE are few things more entertaining to fellow mathematicians than when a famous figure in the field decides, after years of life and research, to reminisce in their presence, and to tell stories about himself and his work. L. J. Mordell, the renowned number theorist and expert in Diphantine analysis, has done this on a number of occasions. The following stories, along with others credited to him in this book, have been culled, with his gracious permission, from his article: "Reminiscences of an octogenarian mathematician," *The American Mathematical Monthly,* Nov. 1971. Though Mordell spent most of his professional life in England, he was born in America—in Philadelphia, in 1888—and lived his first eighteen years there.

123° *A couple of confessions.* The great number theorist, L. J. Mordell, referring to his grammar-school days, said: "I think I was good at arithmetic, but certainly not in later life."

Recalling a mathematics test that he took in 1904, when he was sixteen years old, Mordell said: "All I remember about the examination is that there was a question on Sturm's theorem about equations, which I could not do then and cannot do now."

124° *The beginning of his major interest.* About the age of fourteen, before entering high school, I came across some old algebra books in the five-and-ten-cent counters of Leary's famous bookstore in Philadelphia, and for some strange reason the subject appealed to me. One of these books was *A Treatise on Algebra* by C. W. Hackley, who was a professor from 1843 to 1861 at what was then called Columbia College in New York City. My copy was the third edition, dated 1849 (the first appeared in 1846). It was really a good book, though not rigorous, and contained a great deal of material, including the theory of equations, series, and a chapter on the theory of numbers. Like the old algebra books of those days, it had a chapter on Diophantine analysis, a subject I found most attractive. It is not without interest that in later years much of my best research deals with this subject. In fact, I have recently written a book on the subject,* which appeared in 1969.

—L. J. MORDELL

125° *History repeats itself.* In 1912, an international congress was held in Cambridge, and I attended it. I am fully aware of the implications of the story I am going to tell. I went into the buffet room where all the distinguished mathematicians were gathered, and I thought to myself, "What an odd-looking lot they are." I have no doubt that history repeats itself.—L. J. MORDELL

126° *On a train.* During the Second World War we had a country cottage at Chinley, about twenty miles from Manchester, where I was then a professor. My wife and I were coming home one weekend by train. We entered a compartment, and my wife sat diagonally opposite from me. In front of me was a youth and beside me a middle-aged man. Presently I noticed that the youth was reading a book entitled *Teach Yourself Trigonometry.* Hello, I thought, we are in the same profession. So I asked him whether it was an interesting subject. He did not reply, maintaining a stony silence. Obviously this was an important war secret, and Hitler was not going to get any information from him. Five minutes later I tried again and asked him whether it was a difficult subject. Again no reply, and so I tried no further. When we came to our local station, I got out, and my wife continued into town. She told me

---

*Diophantine Equations. New York: Academic Press, 1969.

afterwards what took place. The other man turned to the youth and said, "You were very rude. Why did you not answer the gentleman?" The reply was, "What does he think he knows about mathematics?"—L. J. MORDELL

127° *Poor sales resistance.*   In 1953, when I was a Visiting Professor at the University of Toronto, I went with my wife to buy a pair of socks. By the time I left the store, she and the salesman persuaded me to buy an overcoat. When I related this to a doctor friend, he said he knew a far better salesman. A woman, whose husband had died, went to buy a suit of clothes to bury him in. The salesman persuaded her to buy a suit with two pairs of trousers. —L. J. MORDELL

128° *The perfect reply.*   In 1958 I was a Visiting Professor at the University of Colorado at Boulder. One day the phone rang, and a woman's voice said, "I am Ann Lee and I want to give you a chance of winning forty-five dollars." I said, "Oh." She then asked me, "What is the oldest dance in the world?" I said, "That's a difficult question and I don't know the answer." She then asked me, "Where does the tango come from?" I replied, "South America." "Good," she said, "you have answered the question, and you are now entitled to forty-five dollars of free dancing lessons." You may think for a moment as to what reply you should make to this, but I would get top marks. I asked her, "Who was the first President of the United States of America?" "George Washington," she said. "Good," I replied, "you have won your forty-five dollars back again."—L. J. MORDELL

129° *Recalling distinguished classmates.*   One of my classmates there [the Grammar School Modell attended in Philadelphia], now Professor F. C. Dietz, is Professor Emeritus of History at the University of Illinois at Urbana; I saw him in December last when I lectured there. I mentioned this to Professor P. Bateman, who is head of the Mathematical Department there, and he countered by saying that one of *his* classmates had been electrocuted for murder.—L. J. MORDELL

130° *Leo Moser's toast to L. J. Mordell.*

Here's a toast to L. J. Mordell,
Young in spirit, most active as well.
   He'll never grow weary
   Of his love, number theory.
The results he obtains are just swell.

FROM OUR OWN TIMES

THE following anecdotes are about mathematicians of our own times. Since most of these people are still living, it is with some trepidation that the stories are inserted, for it certainly is not our intention to injure the feelings of any of our contemporary brethren. Indeed, the inclusion of a story about a living or a recently deceased person, even though the story pokes fun at the person, should be interpreted as a sign of affection for that person. Most of these stories are from a collection begun several years ago by Leo Moser, who forwarded them to Leon Bankoff. After the sad and sudden death of Moser, Bankoff passed the anecdotes on to the present writer, who now offers them, along with a few others, for general perusal. A collection of stories about contemporary mathematicians can easily be expanded to fill a book of its own.

131° *A modest man.* In 1923 I attended a meeting of the American Mathematical Society held at Vassar College in New York State. Someone called Rainich,* from the University of Michigan at Ann Arbor, gave a talk upon the class number of quadratic fields, a subject in which I was then very much interested. I noticed that he made no reference to a rather pretty paper written by one Rabinowitz from Odessa and published in *Crelle's Journal*. I commented upon this. He blushed and stammered and said, "I am

---

*G. Y. Rainich, Emeritus Professor of Mathematics, University of Michigan.

Rabinowitz." He had moved to the United States and changed his name.—L. J. MORDELL

132° *Out of this world.*   One day, while lecturing, Professor Halmos made a mistake at the blackboard. He erased it, saying: "Excuse me. I am always in Hilbert space."—PETER GEDDES

133° *Please accommodate us.*   Of Leo Moser it was said that he was writing a book and taking so long about it that his publishers became very much worried and went to see him. He said he was very sorry about the delay, but he was afraid that the book might have to be a posthumous one. Well, he was told, please hurry up with it.—L. J. MORDELL

134° *Continued fractions.*   Helmut Hasse spoke yesterday on continued fractions. But, of course, he didn't finish.
—CLAYTON W. DODGE

135° *Dan Christie.*   When Professor Dan Edwin Christie of Bowdoin College suddenly passed away on July 18, 1975, some of his close colleagues delivered short talks at the memorial services held four days later in the Bowdoin College Chapel.

Professor James E. Ward III, in his talk, recalled the following little incident. In the preceding summer, Ward taught a course from the manuscript of Dan's last book, and dropped Dan a short note pointing out a minor error. Ward concluded his note by saying, "It's elegant, but it's not correct." Within a matter of hours, Ward received a note back from Dan acknowledging the error, offering an improvement to a correction suggested by Ward, and concluding with, "If it's incorrect, then it's not elegant."

Professor Richard L. Chittim, who presented the final short talk, illustrated Dan's famous one-line retorts by a story reported by Professor Fritz Koelln. Once when Koelln and Christie were talking about childhood homes, and in particular about Milo, Maine, where Dan was born, Koelln laughingly asked Dan if he had known the Venus of Milo. Without hesitation Dan replied, "I went to school with her."

136° *Under the wheels.*   At the Physics Congress held in Zurich in 1931, P. L. Kapitza had himself photographed lying on the ground close to the wheels of a car. He explained, "I just want to know what I should look like if I were being run over."

—LEO MOSER

137° *Pride.*   While still undergraduates, the Austrian Houtermans and the Englishman Atkinson, during a walking tour near Göttingen, began to work out their theory of the thermonuclear reactions on the sun, a theory that later achieved much fame. The theory for the first time put forward the conjecture that solar energy might be attributed not to demolition but to fusion of lightweight atoms. The development of this theory led straight to the hydrogen bomb. At the time (1927), of course, neither of the two young students of the atom dreamed of such sinister consequences.

Houtermans reports: "That evening, after we had finished our essay, I went for a walk with a pretty girl. As soon as it grew dark, the stars came out, one after another, in all their splendor. 'Don't they sparkle beautifully?' cried my companion. But I simply stuck out my chest and said proudly: 'I've known since yesterday why it is that they sparkle.' "—LEO MOSER

138° *Erdös's new suit.*   Paul Erdös came to the University of Syracuse one day in a new suit. His colleagues were surprised. They were even more surprised when Erdös removed his jacket and they noticed that his vest had a rectangular hole cut out of it. "What is the meaning of this?" they inquired. Erdös explained, "There was a ticket sewn to the vest. The ticket just had the size and price of the suit on it and I didn't need these so I cut it out."

—LEO MOSER

139° *Naïveté.*   After seeing the movie *Cyrano de Bergerac* with some friends, Hans Zassenhaus commented, "An excellent picture. The producers were fortunate to find such an excellent actor who at the same time has such a long nose."—LEO MOSER

140° *A good reason.* One cold and stormy day, J. Lambek was on his way to McGill University by taxi. He noticed Zassenhaus on a street corner waiting for a bus, so he had the taxi driver stop and he offered Zassenhaus a lift. "No, thank you," said Zassenhaus. "If I can't get to work on time by bus, then I don't want to get there on time at all."—LEO MOSER

141° *A reasonable excuse.* A. Schild was a graduate student at the University of Toronto. He was working on a thesis on cosmology and, at the same time, attending, or at least supposed to be attending, various lecture courses, including one on topology. He missed many consecutive lectures in this latter course and finally, when he did show up for a lecture, the instructor asked him why he had been absent so long. His answer: "I have been busy calculating the size of the universe."—LEO MOSER

142° *An explanation.* After Loo King Hua returned to Red China, there were (false) reports in some American newspapers that he had committed suicide. R. Ayoub, who was a student of Hua's, explained these reports as follows: "I sent him my Ph.D. thesis and I guess he just couldn't take it."

—LEO MOSER

143° *Progress.* In 1951, E. Teller was asked: "Will the thermonuclear device work?" He replied, "I don't know."

"But if you didn't know five years ago, haven't you made any progress since then?"

"Oh, yes," Teller replied, "Now I don't know on much better grounds."—LEO MOSER

144° *Some new-old results.* J. L. Synge once told A. Weinstein about some new results he had just obtained. Weinstein told Synge that the results were not new and indeed appeared in a recent issue of an Italian journal. Synge said, "That is impossible. I subscribe to this journal and I saw no such article." Weinstein stuck to his opinion, so they went to Synge's office and Synge pulled out the journal in question. Several pages were joined together and when these were separated the paper that Weinstein had mentioned emerged.—LEO MOSER

145° *Poetry and physics.*   One evening Paul Dirac, who was usually so silent, took Oppenheimer aside and gently reproached him. "I hear," he said, "that you write poetry as well as working in physics. How on earth can you do two such things at once? In science one tries to tell people, in such a way as to be understood by everyone, something that no one ever knew before. But in poetry, it's the exact opposite!"—LEO MOSER

146° *Amazing.*   If two integers are chosen at random, the probability that they are relatively prime is $6/\pi^2$. In an article in the *Scientific American,* July 1953, pp. 31–35, entitled "Circumetrics," N. T. Bridgeman describes an experimental verification of this fact as follows:
"Thus Chartres, in 1904, made a random selection of 250 pairs of primes and found that 154 of them were prime to each other; and as it is known that the probability of the conjunction is $6/\pi^2$, his trial amounted to an estimate of π of 3.12."

—LEO MOSER

147° *A noble ambition, but . . . .*   Howard Fehr, on being told by H. S. M. Coxeter that one of Coxeter's children was considering studying for the ministry, remarked, "A noble ambition, but I trust he will grow out of it."—LEO MOSER

148° *Yes, but.*   Leopold Infeld, of the University of Toronto, gave a course in the philosophy of science. At one time he was discussing new concepts introduced into physics in connection with various theories of the origin of the universe. A student asked, "Can one not similarly postulate the existence of God?"
"Of course, of course," said Infeld, "but, my boy, you know quite well that this is not the sensible thing to do."

—LEO MOSER

ON MATHEMATICS AND MATHEMATICIANS

149° *On mathematicians.* The good Christians should beware of mathematicians and all those who make empty prophecies. The danger already exists that the mathematicians have made a covenant with the Devil to darken the spirit and to confine man in the bonds of Hell.—St. AUGUSTINE

150° *An element of luck.* There is an old adage: Oats and beans and barley grow, but neither you nor I nor anybody else knows what makes oats and beans and barley grow. Neither you nor I nor anybody else knows what makes a mathematician tick. It is not a question of cleverness. I know many mathematicians who are far abler and cleverer than I am, but they have not been so lucky. An illustration may be given by considering two miners. One may be an expert geologist, but he does not find the golden nuggets that the ignorant miner does.

—L. J. MORDELL

151° *Existence proofs.* God exists since mathematics is consistent, and the Devil exists since we cannot prove it.

—A. WEIL

152° *The game of mathematics.* God is a child; and when he began to play, he cultivated mathematics. It is the most godly of man's games.

—V. ERATH
Das blinde Spiel (1954)

153° *An analogy.* A student who has merely done mathematical exercises but has never solved a mathematical problem may be likened to a person who has learned the moves of the chess pieces but has never played a game of chess. The real thing in mathematics is to play the game.—STEPHEN J. TURNER

71

154° *Longevity.* Archimedes will be remembered when Aeschylus is forgotten, because languages die and mathematical ideas do not.

—G. H. HARDY
A Mathematician's Apology

155° *The inner self.* In some ways, a mathematician is not responsible for his activities. One sometimes feels there is an inner self occasionally communicating with the outer man. This view is supported by statements made by H. Poincaré and J. Hadamard about their researches.* I remember once walking down St. Andrews Street some three weeks after writing a paper. Though I had never given the matter any thought since then, it suddenly occurred to me that a point in my proof needed looking into. I am very grateful to my inner self for his valuable help in the solution of some important and difficult problems that I could not have done otherwise.—L. J. MORDELL

156° *A felicitous motto.* Ever since issue No. 26, 1901, *The Mathematical Gazette* (the official journal of the Mathematical Association, an association of teachers and students of elementary mathematics in Great Britain) has carried on its cover the apt motto of Roger Bacon:

I hold every man a debtor to his profession, from the which as men of course do seek to receive countenance and profit, so ought they of duty to endeavor themselves by way of amends to be a help and an ornament thereunto.

157° *Mathematics and Antaeus.* The mighty Antaeus was the giant son of Neptune (god of the sea) and Ge (goddess of the earth), and his strength was invincible so long as he remained in contact with his mother earth. Strangers who came to his country were forced to wrestle to the death with him, and so it chanced one day that Hercules and Antaeus came to grips with one another. But

*See Items 348°, 349°, 350°, 351°, 353°, and 358° in *Mathematical Circles Squared.*

Hercules, aware of the source of Antaeus' great strength, lifted and held the giant from the earth and crushed him in the air.

Surely there is a parable here for mathematicians. For just as Antaeus was born of and nurtured by his mother earth, history has shown us that all significant and lasting mathematics is born of and nurtured by the real world. As in the case of Antaeus, so long as mathematics maintains its contact with the real world, it will remain powerful. But should it be lifted too long from the solid ground of its birth into the filmy air of pure abstraction, it runs the risk of weakening. It must of necessity return, at least occasionally, to the real world for renewed strength.

158° *The tree of mathematics.*   [Adapted from Section 15–10 of Howard Eves, *An Introduction to the History of Mathematics,* fourth edition. New York: Holt, Rinehart and Winston, 1976.]

It became popular some years ago to picture mathematics in the form of a tree, usually a great oak tree. The roots of the tree were labeled with such titles as *algebra, plane geometry, trigonometry, analytic geometry,* and *irrational numbers.* From these roots rose the powerful trunk of the tree, on which was printed *calculus.* Then, from the top of the trunk, numerous branches issued and subdivided into smaller branches. These branches bore such titles as *complex variables, real variables, calculus of variations, probability,* and so on, through the various "branches" of higher mathematics.

The purpose of the tree of mathematics was to point out to the student not only how mathematics had historically grown, but also the trail the student should follow in pursuing a study of the subject. Thus, in the schools and perhaps the freshman year at college, he should occupy himself with the fundamental subjects forming the roots of the tree. Then, early in his college career, he should, through a specially heavy program, thoroughly master the calculus. After this is accomplished, the student can then ascend those advanced branches of the subject that he may wish to pursue.

The pedagogical principle advocated by the tree of mathematics is probably a sound one, for it is based on the famous law pithily stated by biologists in the form: "Ontogeny recapitulates phylogeny," which simply means that, in general, "The individual

73

repeats the development of the group." That is, at least in rough outline, a student learns a subject pretty much in the order in which the subject developed over the ages. As a specific example, consider geometry. The earliest geometry may be called *subconscious geometry*, which originated in simple observations stemming from human ability to recognize physical form and to compare shapes and sizes. Geometry then became *scientific*, or *experimental*, *geometry*, and this phase of the subject arose when human intelligence was able to extract from a set of concrete geometrical relationships a general abstract relationship (a geometrical law) containing the former as particular cases. The bulk of pre-Hellenic geometry was of this experimental kind. Later, actually in the Greek period, geometry advanced to a higher stage and became *demonstrative geometry*. The basic pedagogical principle here under consideration claims, then, that geometry should first be presented to young children in its subconscious form, probably through simple artwork and simple observations of nature. Then, somewhat later, this subconscious basis is evolved into scientific geometry, wherein the pupils induce a considerable array of geometrical facts through experimentation with compasses and straightedge, with ruler and protractor, and with scissors and paste. Still later, when the student has become sufficiently sophisticated, geometry can be presented in its demonstrative, or deductive, form, and the advantages and disadvantages of the earlier inductive processes can be pointed out.

So we have here no quarrel with the pedagogical principle advocated by the tree of mathematics. But what about the tree itself? Does it still present a reasonably true picture of present-day mathematics? We think not. A tree of mathematics is clearly a function of time. The oak tree described above certainly could not, for example, have been the tree of mathematics during the great Alexandrian period. The oak tree does represent fairly well the situation in mathematics in the eighteenth century and a good part of the nineteenth century, for in those years the chief mathematical endeavors were the development, extension, and application of the calculus. But with the enormous growth of mathematics in the twentieth century, the general picture of mathematics as given by

74

the oak tree no longer holds. It is perhaps quite correct to say that today the larger part of mathematics has no, or very little, connection with the calculus and its extensions. Consider, for example, the vast areas covered by abstract algebra, finite mathematics, set theory, combinatorics, mathematical logic, axiomatics, nonanalytical number theory, postulational studies of geometry, finite geometries, and on and on.

We must redraw the tree of mathematics if it is to represent mathematics of today. Fortunately there is an ideal tree for this new representation—the banyan tree. A banyan tree is a many-trunked tree, ever growing newer and newer trunks. Thus, from a branch of a banyan tree, a threadlike growth extends itself downward until it reaches the ground. There it takes root and over the succeeding years the thread becomes thicker and stronger, and in time becomes itself a trunk with many branches, each dropping threadlike growths to the ground.

There are some banyan trees in the world having many scores of trunks, and covering many city blocks in area. Like the great oak tree, these trees are both beautiful and long-lived; it is claimed that the banyan tree in India, against which Buddha rested while meditating, is still living and growing. We have, then, in the banyan tree a worthy and more accurate tree of mathematics for today. Over future years newer trunks will emerge, and some of the older trunks may atrophy and die away. Different students can select different trunks of the tree to ascend, each student first studying the foundations covered by the roots of his chosen trunk. All these trunks, of course, are connected overhead by the intricate branch system of the tree. The calculus trunk is still alive and doing well, but there is also, for example, a linear algebra trunk, a mathematical logic trunk, and others.

Mathematics has become so extensive that today one can be a very productive and creative mathematician and yet have scarcely any knowledge of the calculus and its extensions. We who teach mathematics in the colleges today are probably doing a disservice to some of our mathematics students by insisting that *all* students must first ascend the calculus trunk of the tree of mathematics. In spite of the great fascination and beauty of calculus, not all stu-

75

dents of mathematics find it their "cup of tea." By forcing *all* students up the calculus trunk, we may well be killing off some potentially able mathematicians of the noncalculus fields. In short, it is perhaps time to adjust our mathematical pedagogy to fit a tree of mathematics that better reflects the recent historical development of the subject.

159° *Another analogy.*   The game of chess has always fascinated mathematicians, and there is reason to suppose that the possession of great powers of playing that game is in many features very much like the possession of great mathematical ability. There are the different pieces to learn, the pawns, the knights, the bishops, the castles, and the queen and the king. The board possesses certain possible combinations of squares, as in rows, diagonals, etc. The pieces are subject to certain rules by which their motions are governed, and there are other rules governing the players . . . One has only to increase the number of pieces, to enlarge the field of the board, and to produce new rules which are to govern either the pieces or the player, to have a pretty good idea of what mathematics consists.—J. B. SHAW

160° *Stepping down.*   Occasionally a college professor is appointed to the position of a deanship, and thenceforth no longer does anything academically productive. This has happened frequently to mathematics professors, and one is reminded of the following perhaps apocryphal story.

A mathematics professor developed a brain tumor and had to undergo an emergency operation. The surgeon detached the professor's cranium, took out the brain, and laid it on the table.

Right then a colleague of the professor arrived and announced that the professor had just been appointed to a deanship. With a whoop of joy, the professor bounced up from the operating table, slapped on his cranium, and dashed for the nearest exit. "Wait," cried the surgeon, "you've forgotten to put back your brain."

"I won't need it now," called back the professor over his shoulder, "I'm a dean."

161° *A least upper bound.* S. Jennings, of the University of British Columbia, had become increasingly active in university administration. He was characterized by a colleague as "a least upper bound of non-deans."—LEO MOSER

162° *A great mistake.* In 1688 Cambridge University selected Isaac Newton as their representative in Parliament. Not a very good choice, it would seem. During his entire tenure in Parliament, Newton's only known speech was a request to have a window opened.

PROFESSORS, TEACHERS, AND STUDENTS

163° *Julian Lowell Coolidge.* Julian Lowell Coolidge, the great geometer at Harvard in the first half of the twentieth century, was a wit and a humorist in class. "I definitely try," it is said he once remarked, "when I teach, to make the students laugh. And while their mouths are open, I put something in for them to chew on."

164° *A special course.* A mathematics professor was asked by his dean to prepare a special make-up course for a group of sick students at the university. The professor labeled the course, "Mathematics for Ill Literates."

165° *For the lazy student.* Bennett Cerf has reported that Frank Boyden, a famous headmaster of Deerfield Academy, kept the following little poem on hand for lazy students:

> You can't go far just by wishing
> Nor by sitting around to wait.
> The good Lord provided the fishing—
> But you have to dig the bait.

One is reminded of the little jingle given in Item 190° of *Mathematical Circles Revisited.*

166° *A ticklish situation.* A mathematics professor forgot an important afternoon appointment at his campus office that he had made with one of his students. After waiting at the professor's office for over a half hour, the student, with some trepidation, got up the courage to go to the professor's home in town. When he rang the bell, the door was answered by the professor's small son.

"Where can I get hold of your father?" inquired the nervous student.

"I wouldn't know," replied the little boy, "He's ticklish all over."

167° *Mars.* The professor of mathematical astronomy, wishing to demonstrate something about the relative positions of some of the planets, said, "I will let my hat here represent the planet Mars. Any questions?"

"Yes," replied a student, "Is Mars inhabited?"

168° *Math exams.* "A fool," sighed the annoyed mathematics professor, "can ask more questions in a few minutes than a wise man can answer in hours." At that, one of his students was heard to murmur in a barely audible voice, "No wonder so many of us flunked your last exam."

169° *Multiple use.* On the wall in the hallway just outside the mathematics department lounge there is a row of hooks along with a sign reading, "For Faculty Members Only." Some witty student added in pencil below, "May also be used for hats and coats."

170° *How's that!* A school board advertised for a teacher of algebra and geometry. One of the replies read: "Gentlemen, I noticed your advertisement for a teacher of algebra and geometry, male or female. Having been both for several years, I offer you my services."—LEO MOSER

171° *Research versus teaching.* A King once decided to honor that one of his subjects who had contributed the most to the furtherance of knowledge. Forthwith appeared a number of top-flight researchers in the various fields of study. After carefully considering the credentials of the candidates, the King noticed a stooped and shabbily dressed old woman standing in the background.

"Who is that woman?" asked the King.

"She has come merely to observe, Sire," replied the King's minister. "You see, she is interested in the outcome because she taught all these candidates when they were young."

The King descended from his throne and placed the wreath of honor on the old teacher's brow.

172° *Researchers and teachers.* Edith Wharton has pointed out, "There are two ways of spreading light: to be the candle, or the mirror that reflects it."

LECTURES

A LARGE part of a mathematics professor's life is spent in lecturing, either before college classes or at special mathematical gatherings. Sometimes amusing incidents occur during, or relative to, these lectures. Here are eight such stories gathered by Leo Moser. Note the similarity of the last three stories, wherein the existence of a *proof* of a mathematical result renders *checking* the result both mundane and extraneous.

173° *An understanding evaluation.* At the end of an hour lecture to a large audience at a mathematical congress in New York, C. S. Peirce remarked, "Come to think of it, there is only one person who might have had some chance of understanding the gist of my lecture today—and he is in South America."

—LEO MOSER

174° *A crackpot.*   Jacques Hadamard once planned a lecture trip in the United States. He wanted to meet American mathematicians and speak to many of the mathematics faculties concerning mathematical problems. He wrote, among other places, to Syracuse University. Since the head of the department was away, the letter reached the president of the university. The president refused Hadamard's request to have a lecture arranged for him. He argued that since Hadamard had not mentioned monetary remuneration, he must be a crackpot.—LEO MOSER

175° *Understanding.*   One year L. Infeld and R. Brauer gave a joint course on group theory and quantum mechanics. According to the graduate students present, Brauer didn't understand Infeld, Infeld didn't understand Brauer, and the students, of course, didn't understand either one.—LEO MOSER

176° *Interruption.*   Antoni Zygmund was constantly annoyed by a young female student who often interrupted his lecture with, "I don't understand this." Finally Zygmund said to her, "Don't worry about it. You are still so young."—LEO MOSER

177° *An uneasy start.*   Professor K. D. Fryer, of the University of Waterloo, was to give a lecture on continuous geometry at Queens University. Before the lecture, he filled the blackboard with very complicated-looking formulas and left a large note over these formulas: DO NOT ERASE. When the audience arrived they were rather worried by the complexity of the formulas and felt that the lecturer would soon leave them far behind. Fryer began the lecture with, "Well, I am sure we can dispense with all this." And he erased all the formulas.—LEO MOSER

178° *Checking.*   In a mathematics colloquium lecture at the University of Alberta, Max Wyman gave a long and complicated proof of a theorem which, once proved, could be checked quite easily. P. G. Rooney asked Wyman how the check came out and Wyman replied, "I didn't check it. I didn't have to. I proved it!" —LEO MOSER

179° *Bier Bauch.*   L. Bieberbach was overly fond of beer and as a result had a pot belly. He was known to his students as Bier Bauch (beer belly).

Bieberbach often had considerable difficulty with elementary arithmetic. His students knew his failing, so whenever he completed the proof of a theorem that could be illustrated by a numerical example, they would ask him to give such an example. After several unsuccessful attempts, he would give up in exasperation and exclaim, "Es ist doch so wie so bewiessen [In any case it has been proved]."—LEO MOSER

180° *Artin on graphs.*   Emil Artin gave a lecture on graphs at the University of Toronto. The problem he treated and solved was suggested by the question of how many chemical compounds satisfying certain valency conditions are possible. After the lecture, a chemist asked Artin how his result compared with the actual number of compounds. Artin replied, "I don't know and I don't care."—LEO MOSER

QUADRANT THREE

From author's jokes
to a difficult problem

AUTHORS AND BOOKS

SOMETIMES an author of a mathematics text will indulge in a literary prank or spring a joke of significance only to the brethren. And then again, sometimes a nonmathematical author will attempt a bit of mathematical discussion that a mathematician may find interesting or amusing. Here are some examples.*

181° *Authors' jokes.*   Leonard Gillman and Robert McDowell, in their book *Calculus* (W. W. Norton, 1973), include, "for the sake of completeness," a theorem concerning greatest lower bounds. Elsewhere in the book they state that "a little reflection shows that the graph of f^{-1} is the mirror image . . . of the graph of f."

182° *The Chinese Remainder Theorem.*   On p. 72 of Harold M. Stark's *An Introduction to Number Theory* appears, in connection with the Chinese Remainder Theorem, the following cute footnote: "More rarely known as the Formosa Theorem."

183° *Mohr–Mascheroni constructions.*   Commenting on Mohr–Mascheroni constructions (constructions utilizing the compasses only), M. H. Greenblatt, in his book *Mathematical Entertainments,* says: "Mohr of these Mascheroni constructions are described in a booklet by A. N. Kostovskii."

184° *Honoring a friend.*   There is a story that says when the witty Paul Halmos wrote his little gem, *Finite-Dimensional Vector Spaces,* he promised a friend that he would mention the friend's name in the book. True enough, in the index of the book one finds the name, "Hochschild, G. P." The page reference is 198, which

---

*Earlier examples may be found in Item 192° of *Mathematical Circles Revisited* and Items 344°, 345°, 346°, and 347° of *Mathematical Circles Squared.*

turns out to be the number of the page in the index where the name is mentioned.

Hochschild later became an eminent professor of mathematics at the University of California at Berkeley.

185° *Close.*   In the book *Introduction to Modern Prime Number Theory,* a certain theorem is proved by the indirect method. The contradiction comes out in the form $333 \geq 1000/3$. The author, T. Estermann, remarks: "The narrow margin by which this contradiction was obtained reminds me of the story of the Scotsman who looked suspiciously at his change, and when asked if it was not enough, said: 'Yes, but only just.' "—LEO MOSER

186° *Among the authors.*   Helmut Hasse, in the index of authors at the rear of his *Vorlesungen über Zahlentheorie* (1950), lists "Gott," with a reference to page 1. On page 1 one finds quoted Kronecker's famous remark, "Die ganzen Zahlen hat Gott gemacht, alles andere ist Menschenwerk."

187° *A deletion.*   In my Springer textbook *Vorlesungen über Zahlentheorie,* at the very beginning, I mention Kronecker's famous dictum and Dedekind's quite opposite opinion. In 1953 a Russian translation of this book appeared in Moscow, and in it the first paragraph is left out. When I went to Moscow in 1963, they told me there that this paragraph was not admitted by the State Editors of Foreign Literature, because God does not exist.

—HELMUT HASSE

188° *Whewell's poetry.*   Dr. William Whewell, when Master of Trinity College, much to the amusement of his students, unintentionally rhymed some of the prose in the first edition (1819) of his *An Elementary Treatise on Mechanics.* He wrote: "There is no force, however great, can stretch a cord, however fine, into a horizontal line, which is accurately straight," which forms the tetrastich:

There is no force, however great,
Can stretch a cord, however fine,
Into a horizontal line,
Which is accurately straight.

The accidental rhyme was first brought to Whewell's attention when Professor Sedgwick of the geology department at Cambridge recited the lines in an after-dinner speech. Whewell, who failed to see any humor in the matter, altered the lines in the following edition of his work so as to eliminate the poem. Ironically, during his lifetime Whewell published two books of serious poetry, but the lines quoted above constitute the only "poem" by him remembered today.

189° *A criticism.* David Widder, of Harvard University, was once asked, "What is your opinion of Bateman's book on differential equations? It is a very good book, is it not?" "That book!" Widder exclaimed, "Open it to any page at random and show me any statement in heavy print and I will give you a counterexample."—LEO MOSER

190° *Proving a postulate.* At a social gathering given by the von Rinnlingens in Thomas Mann's famous short story *Little Herr Friedemann,* a number of the guests retire to the smoking-room. We read:

> Some of the men stood talking in this room, and at the right of the door a little knot had formed round a small table, the center of which was the mathematics student, who was eagerly talking. He had made the assertion that one could draw through a given point more than one parallel to a straight line; Frau Hagenström had cried out that this was impossible, and he had gone on to prove it so conclusively that his hearers were constrained to behave as though they understood.

Thomas Mann was born in Lübeck, Germany, in 1875, and died in Zurich, Switzerland, in 1955. *Little Herr Friedemann* was his first literary success; he wrote it in his early twenties while on a two-year stay in Italy.

191° *A classification of minds.*   I was just going to say, when I was interrupted, that one of the many ways of classifying minds is under the heads of arithmetical and algebraical intellects. All economical and practical wisdom is an extension of the following arithmetical formula: $2 + 2 = 4$. Every philosophical proposition has the more general character of the expression $a + b = c$. We are mere operatives, empirics, and egotists until we learn to think in letters instead of figures.

—OLIVER WENDELL HOLMES
The Autocrat of the Breakfast Table

192° *About vectors.*   It is rumored that there was once a tribe of Indians who believed that arrows are vectors. To shoot a deer due northeast, they did not aim an arrow in the northeasterly direction; they sent two arrows simultaneously, one due north and the other due east, relying on the powerful resultant of the two arrows to kill the deer.

Skeptical scientists have doubted the truth of this rumor, pointing out that not the slightest trace of the tribe has ever been found. But the complete disappearance of the tribe through starvation is precisely what one would expect under the circumstances; and since the theory that the tribe existed confirms two such diverse things as the NONVECTORIAL BEHAVIOR OF ARROWS and the DARWINIAN PRINCIPLE OF NATURAL SELECTION, it is surely not a theory to be dismissed lightly.

—BANESH HOFFMAN
About Vectors (Prentice-Hall, Inc., 1966)

193° *Asymptotic approach.*   . . . he seemed to approach the grave as an hyperbolic curve approaches a line—less directly as he got nearer, till it was doubtful if he would ever reach it at all.

THOMAS HARDY
Far from the Madding Crowd
about the malster, in Chapter XV

194° *A literary allusion.* James Hilton, in his haunting novel *Lost Horizon,* describes Conway's first impression of the foreground leading to Shangri-La, and of the more remote, conical, snow-covered Mt. Karakal that tenuously towers over Shangri-La.

> It was not a friendly picture, but to Conway, as he surveyed, there came a queer perception of fineness in it, of something that had no romantic appeal at all, but a steely, almost intellectual quality. The white pyramid in the distance compelled the mind's assent as passionlessly as a Euclidean theorem . . . .

Later in the story we read of the interplay of the atmosphere and the mystery of Shangri-La on Conway:

> Its atmosphere soothed while its mystery stimulated, and the total sensation was agreeable. For some days now he had been reaching, gradually and tentatively, a curious conclusion about the lamasery and its inhabitants; his brain was still busy with it, though in a deeper sense he was unperturbed. He was like a mathematician with an abstruse problem—worrying over it, but worrying very calmly and impersonally.

At still another point in the story, Conway notes his reaction to the discourses of the High Lama:

> At times he had the sensation of being completely bewitched by the mastery of that central intelligence, and then, over the little pale blue tea-bowls, the celebration would contract into a liveliness so gentle and miniature that he had an impression of a theorem dissolving limpidly into a sonnet.

One is reminded of Sherlock Holmes' disparaging comment to Watson on the completion of the latter's first book, *A Study in Scarlet*:

> Honestly, I cannot congratulate you on it. Detection is, or ought to be, an exact science, and should be treated in the same cold and unemotional manner. You have attempted to tinge it with romanticism, which produces the same effect as if you worked a love story or elopement into the 5th proposition of Euclid.

DEFINITIONS

THE forming of accurate definitions is of great importance in mathematics.

195° *A definition.*   A little girl was asked for her definition of *nothing.* *"Nothing,"* she replied, "is like a balloon with its skin off."

196° *On definitions.*   Making good definitions is not easy, The story goes that when the philosopher Plato defined *man* as "a two-legged animal without feathers," Diogenes produced a plucked cock and said, "Here is Plato's man." Because of this, the definition was patched up by adding the phrase "and having broad nails"; and there, unfortunately, the story ends. But what if Diogenes had countered by presenting Plato with the feathers he had plucked?

—BANESH HOFFMAN
About Vectors (Prentice-Hall, Inc., 1966)

197° *A suggestion.*   A short word to replace the often-used phrase "transformation of coordinates" would be handy in geometry. How about "code"?

198° *Clayton Dodge's rectangle (wrecked angle).*

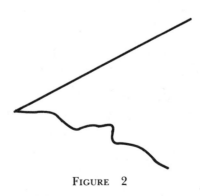

FIGURE   2

199° *Cute.*   Question: Define *acute right triangle.* Answer: A right triangle with pretty legs.—Clayton W. Dodge

200° *A Fourier series.*   Yea + Yea + Yea + Yea.
—Clayton W. Dodge

201° *Yes.*   On his preliminary Ph.D. oral examination, W. Crawford was asked to explain the meaning of the terms *homeomorphism* and *homomorphism.* He knew what a homeomorphism is, but couldn't quite remember the definition of a homomorphism. He nevertheless concocted a correct answer, namely: "A homomorphism is an isomorphism that didn't quite make it."
—Leo Moser

LOGIC

A mathematical theory results from the interplay of two factors, a set of postulates and a logic. The set of postulates constitutes the basis from which the theory starts and the logic constitutes the rules by which such a basis may be expanded into a body of theorems. [From Howard Eves and C. V. Newsom, *An Introduction to the Foundations and Fundamental Concepts of Mathematics,* revised edition. Holt, Rinehart and Winston, 1965.]

202° *Cause and effect.*   It is well known that for every cause there is an effect and for every effect there is a cause, but sometimes the two are so inextricably tangled together that it is difficult to tell which is which. Consider, for example, the case of the two desert nomads visiting Wisconsin for their first time and seeing a speedboat pulling a zig-zagging water skier about a lake. "Why does the boat go so fast?" inquired one of the nomads. Replied the other, "Because it is chased by a lunatic on a string."

203° *Deduction.*   Some postal authorities are amazingly persevering and ingenious. It is said that one such official came upon a letter addressed to:

91

Wood.
Mr.
Mass.

and promptly and correctly had it delivered to:

Mr. Underwood
Andover,
Mass.

204° *No wonder.*   Two Irish ladies argued with one another every day from their windows, across an intervening lane. They never agreed, which was to be expected of course, for they argued from different premises.

205° *Benchley's law of dichotomy.*   Robert Benchley remarked, "There may be said to be two classes of people in the world: those who constantly divide the people of the world into two classes and those who do not."

206° *So it seems.*   While watching a TV newscast, a woman turned toward her husband and observed: "It seems to me that the majority of people in this country belong to some minority group."

207° *No doubt.*   A reporter asked Undersecretary of State Robert J. McCloskey to comment on the art of decision-making. The Undersecretary replied: "One thing you must keep in mind here: that all decisions aren't made until all decisions are made."

208° *Logical.*   Mayor Richard J. Daly of Chicago, when asked by a reporter to comment on the current trucking strike, replied: "What keeps people apart is their inability to get together."

209° *Necessary and sufficient.*   A local radio announcer, commenting on hazardous driving conditions, advised: "Please don't do any unnecessary driving unless it's absolutely necessary."

210° *Come again.*   An auto-accident report to an insurance company contained the following statement: "My car sustained no damage whatever and the other car somewhat less."

211° *Separate or together.*   Finkbeiner and I were having breakfast in San Antonio and asked for the check. The waitress said, "You want them separate or together?" I said, "Separate, please." She turned to Finkbeiner and said, "You want yours separate too?"—RALPH BOAS

212° *Who's who?*   I met Tom Jones and I said, "How have you been, Jones?" And Jones replied, "Fair to middling, thank you. How have you been, Smith?" "Smith," I said, "That's not my name!" "Nor is my name Jones," said the other fellow. Then we looked each other over again, and true! It was neither of us!

213° *Proof.*   Ambling Andy was telling a group about his great love of walking. He said that as he kept extending his daily walks, he found himself farther and farther from home each day when evening approached. He solved the difficulty by moving to a house halfway up a big conical hill, and thenceforth he spent his days walking round and round the side of the hill, starting each day with the rising sun behind him and ending each day with the setting sun before him. After some years of this he made the tragic discovery that by continual walking around the conical hill he had worn his uphill leg shorter than his downhill leg, and his friends commiserated that the walking days of Ambling Andy were essentially over. But, no. Andy merely reversed his direction of walking around the hill until he wore his long leg down to the length of his short leg. When a listener appeared incredulous, Andy proved the whole matter by standing before the disbelieving person and pointing out: "See, aren't my two legs the same length?"

214° *Of course.*   Question: How do you pronounce the word "ghot"?
Answer: As "fish," by taking the "gh" as in "laugh," the "o" as in "women," and the "t" as in "nation."

93

215° *A syllogism.*

One cat has one more tail than no cat.
No cat has two tails.

Therefore, one cat has three tails.

216° *Double negative.*   A five-year old was being cajoled to recite a poem. To prod him, his mother said, "You just don't know, do you?" To this the youngster replied, "I do not don't know!"

217° *Naturally.*   Grandma can never find her glasses anymore—so she drinks from the can.

218° *An exception.*   An exception is never used to prove a rule, but rather to test a rule. That is, if a rule works well in exceptional (rather than simple) cases, the chances are excellent that it is a good rule.

219° *Ability versus nonability.*

(1) I love mathematics; I attend the meetings of the Mathematics Club and I study the subject a great deal. Still I can't do well in mathematics courses.

(1) I love baseball; I watch it on TV and I study books on the subject. Still I'm a pretty awful player.

(2) I study mathematics much harder than Joe does, but he still gets much better grades.

(2) I practice wrestling much more than Joe does, but he still always beats me.

220° *A law of physics.*   In a little New England town, a brass band of a dozen musicians was blaring away, under the energetic leadership of a puffing and uniformed maestro, before one of the houses of the town.

A curious passing tourist stopped and inquired of an onlooker, "Who are they serenading?"

"Oh, the mayor," was the reply. "Why doesn't he come to the

door or a window and acknowledge the compliment?" continued the tourist.

"Because that's the mayor leading the band," explained the onlooker. "You can't expect him to be in two different places at the same time, can you?"

221° *Language difficulties.*   A Japanese mathematician wrote to an American mathematician requesting a reprint. He concluded with: "And above all, please excuse my shameless desire."

An Indian student wrote the head of the mathematics department of a Canadian university and expressed interest in continuing his studies in Canada. His financial condition apparently was not good, so he inquired, "I wonder what you can do to help me make both my ends meet."—Leo Moser

ON MATHEMATICS AND LOGIC

Many eminent, and some not so eminent, persons have commented on the relative positions of logic and mathematics.

222° *Bacon's misconceived notion.*   It has come to pass, I know not how, that Mathematics and Logic, which ought to be but the handmaids of Physic, nevertheless presume on the strength of the certainty which they possess to exercise dominion over it.
—Francis Bacon
De Augmentis

223° *Successive approximation.*   It is commonly considered that mathematics owes its certainty to its reliance on the immutable principles of formal logic. This . . . is only half the truth imperfectly expressed. The other half would be that the principles of formal logic owe such a degree of permanence as they have largely to the fact that they have been tempered by long and varied use by mathematicians. "A vicious circle!" you will perhaps say. I should rather describe it as an example of the process known by mathematicians as the method of successive approximation.—Maxime Bôcher

224° *Which is which?*   George Bruce Halsted has described mathematics as the "giant pincers" of logic. Isn't it equally true that logic is the "giant pincers" of mathematics?

225° *An identity.*   . . . the two great components of the critical movement, though distinct in origin and following separate paths, are found to converge at last in the thesis: Symbolic Logic is Mathematics, Mathematics is Symbolic Logic, the twain are one. —CASSIUS J. KEYSER

226° *Monadology.*   It seemed to Leibniz that if all the complex and apparently disconnected ideas which make up our knowledge could be analysed into their simple elements, and if these elements could each be represented by a definite sign, we should have a kind of "alphabet of human thoughts." By the combination of these signs (letters of the alphabet of thought) a system of true knowledge would be built up, in which reality would be more and more adequately represented or symbolized. . . . Thus it seemed to Leibniz that a synthetic calculus, based upon a thorough analysis, would be the most effective instrument of knowledge that could be devised. "I feel," he says, "that controversies can never be finished, nor silence imposed upon the Sects, unless we give up complicated reasonings in favor of simple *calculations,* words of vague and uncertain meaning in favor of fixed symbols." Thus it will appear that "every paralogism is nothing but *an error of calculation.*" "When controversies arise, there will be no more necessity of disputation between two philosophers than between two accountants. Nothing will be needed but that they should take pen in hand, sit down with their counting-tables, and (having summoned a friend, if they like) say to one another: *Let us calculate.*" —ROBERT LATTA

227° *The foundation for a scientific education.*   Formal thought, consciously recognized as such, is the means of all exact knowledge; and a correct understanding of the main formal sciences, Logic and Mathematics, is the proper and only safe foundation for a scientific education.—ARTHUR LEFEVRE

228° *The identity again.* Pure mathematics was discovered by Boole in a work he called 'The Laws of Thought' . . . His work was concerned with formal logic, and this is the same thing as mathematics.—BERTRAND RUSSELL

229° *And again.* Mathematics is but the higher development of Symbolic Logic.—W. C. D. WHETHAM

230° *Symbolic Logic.* Symbolic Logic has been disowned by many logicians on the plea that its interest is mathematical, and by many mathematicians on the plea that its interest is logical.
—ALFRED NORTH WHITEHEAD

231° *New logics.* [This, and the succeeding, items are adapted from Howard Eves and C. V. Newsom, *An Introduction to the Foundations and Fundamental Concepts of Mathematics,* revised edition. Holt, Rinehart and Winston, 1965.]

An interesting analogy (if it is not pushed too far) exists between the parallelogram law of forces and the postulational method. By the parallelogram law, two component forces are combined into a single resultant force. Different resultant forces are obtained by varying one or both of the component forces, although it is possible to obtain the same resultant force by taking different pairs of initial component forces. Now, just as the resultant force is determined by the two initial component forces, so (see Figure 3) is a mathematical theory determined by a set of postu-

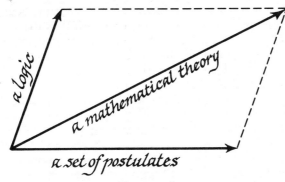

FIGURE 3

97

lates and a logic. That is, the set of statements constituting a mathematical theory results from the interplay of an initial set of statements, called the postulates, and another initial set of statements, called the logic or the rules of procedure. For some time mathematicians have been aware of the variability of the first set of initial statements, namely, the postulates, but until very recent times the second set of initial statements, namely, the logic, was universally thought to be fixed, absolute, and immutable. Indeed, this is still the prevailing view among most people, for it seems quite inconceivable, except to the very few students of the subject, that there could be any alternative to the laws of logic stated by Aristotle in the fourth century B.C. The general feeling is that these laws are in some way attributes of the structure of the universe and that they are inherent in the very nature of human reasoning. As with many other absolutes of the past, this one, too, has toppled, but only as late as 1921. The modern viewpoint can hardly be more neatly put than in the following words of the outstanding American logician, Alonzo Church.

We do not attach any character of uniqueness or absolute truth to any particular system of logic. The entities of formal logic are abstractions, invented because of their use in describing and systematizing facts of experience or observation, and their properties, determined in rough outline by this intended use, depend for their exact character on the arbitrary choice of the inventor. We may draw the analogy of a three-dimensional geometry used in describing physical space, a case for which, we believe, the presence of such a situation is more commonly recognized. The entities of the geometry are clearly of abstract character, numbering as they do planes without thickness and points which cover no area in the plane, point sets containing an infinitude of points, lines of infinite length, and other things which cannot be reproduced in any physical experiment. Nevertheless the geometry can be applied to physical space in such a way that an extremely useful correspondence is set up between the theorems of the geometry and observable facts about material bodies in space. In building the geometry, the proposed application to physical space serves as a rough guide in determining what properties the abstract entities shall have, but does not assign these properties completely. Consequently there may be, and actually are, more than one geometry whose use is feasible in describing physical space. Similarly, there exist, undoubtedly, more than one formal system whose use as a logic is feasible, and of these systems one may be more pleasing or more convenient than another, but it cannot be said that one is right and the other wrong.

232° *A brief history of the new logics.*   New geometries first came about through the denial of Euclid's parallel postulate, and new algebras first came about through the denial of the commutative law of multiplication. In a similar fashion, the new so-called "many-valued" logics first came about by denying Aristotle's law of excluded middle. According to this law, the disjunctive proposition $p \lor$ not-p is a tautology, and a proposition p in Aristotelian logic is always either true or false. Because a proposition may possess any one of two possible truth values, namely truth or falsity, this logic is known as a two-valued logic. In 1921, in a short two-page paper, J. Lukasiewicz considered a three-valued logic, or a logic in which a proposition p may possess any one of three possible truth values. Very shortly after, and independently of Lukasiewicz's work, E. L. Post considered m-valued logics, in which a proposition p may possess any one of m possible truth values, where m is an integer greater than 1. If m exceeds 2, the logic is said to be *many-valued.* Another study of m-valued logics was given in 1930 by Lukasiewicz and A. Tarski. Then, in 1932, the m-valued truth systems were extended by H. Reichenbach to an infinite-valued logic, in which a proposition p may assume any one of infinitely many possible values.

Not all new logics are of the type just discussed. Thus A. Heyting has developed a symbolic two-valued logic to serve the intuitionist school of mathematicians; it differs from Aristotelian logic in that it does not universally accept the law of excluded middle or the law of double negation. Like the many-valued logics, then, this special-purpose logic exhibits differences from Aristotelian laws. Such logics are known as *non-Aristotelian logics.*

Like the non-Euclidean geometries, the non-Aristotelian logics have proved not to be barren of application. Reichenbach actually devised his infinite-valued logic to serve as a basis for the mathematical theory of probability. And in 1933 F. Zwicky observed that many-valued logics can be applied to the quantum theory of modern physics. Many of the details of such an application have been supplied by Garrett Birkhoff, J. von Neumann, and H. Reichenbach. Lukasiewicz has employed three-valued logics to establish the independence of the postulates of familiar two-valued logic. The part that non-Aristotelian logics may play in the future development of mathematics is uncertain but intriguing to contemplate; the application of Heyting's symbolic logic to intuitionist mathematics indicates that the new logics may be mathematically valuable.

COUNTING

THE basis of counting is the notion of a one-to-one correspondence.

233° *A one-to-one correspondence.* One of the earliest literary references to the primitive method of keeping a count by setting up a one-to-one correspondence occurs in the Homeric legends about Ulysses. When Ulysses left the land of the Cyclops, after blinding the one-eyed giant Polyphemus, it is narrated that that unfortunate old giant would sit each morning near the entrance to his cave and from a heap of pebbles would pick up one for each ewe that he let pass out of the cave. Then, in the evening, when the ewes returned, he would drop one pebble for each ewe that he admitted to the cave. In this way, by exhausting the supply of pebbles he had picked up in the morning, he was assured that all his flock had returned in the evening.

234° *Another one-to-one correspondence.* Like the famous David Hilbert in his older age, there was a retired professor of mathematics who was quite a devil with the ladies, still charming the daylights out of them at seventy-seven. In fact, on his seventy-seventh birthday, the professor decided to cut a notch in his cane for each of his new conquests. And that's what killed him on his seventy-eight birthday; he made the mistake of leaning on his cane.

235° *Buttons.* Carrier: "Can 'ee spell?"
"Yes!"
Carrier: "Cipher?"
"Yes!"
Carrier: "That's more than I can; I counts upon my fingers. When they be used up, I begin upon my buttons. I han't got no buttons—visible, that is—'pon my week-a-day clothes, so I keeps the long sums for Sundays, and adds 'em up and down my weskit during sermon. Don't tell any person."
"I won't."
Carrier: "That's right; I don't want it known. Ever seen a gypsy?"
"Oh, yes, often."
Carrier: "Next time you see one, you'll know why he wears so many buttons. You've a lot to learn."

—SIR ARTHUR QUILLER-COUCH
The Ship of Stars

NUMBERS

HERE are still more stories—some silly, some serious—about numbers.

236° *Nein.* In view of the recent revealing of hanky-panky conducted by a number of our government officials in Washington, D.C., an attractive German fraulein in that city was questioned if she had ever been away on trips with senators. She indignantly replied, *"Nein"*—and so was deported.

237° *Round letters.* And then there was the boss who complained that his secretary spelled like a mathematician—she rounded her words off to the nearest letter of the alphabet.

238° *Experimental error.* An experimental physicist was testing the conjecture that all odd numbers are prime. He started by saying, "Three—that's prime. Five—that's prime, too. Seven—

still prime. Nine—oops! What's wrong? But let's go on a bit further. Eleven—that's prime. Thirteen—still prime. Nine must have been an experimental error!"—M. H. GREENBLATT

239° *The psychology of numbers.*   There was a famous Betty Crocker packaged cake mix where, on the box, the busy cook was instructed to add to the contents of the package one egg and half a cup of water *less* one tablespoon, mix for two minutes at medium speed with an electric mixer, pour into a greased 9-inch pan, and then bake for 35 minutes in a moderate (350°F) oven.

Now the cake would bake just as well using a full half-cup of water, but interviews and consumer research revealed that then the recipe would appear too simple, and would leave the baker with guilt feelings. That is, if the cake batter should be too easy to prepare, the cook's basic honesty would prevent her from believing that she had made the cake herself, and her pride in the finished cake would be destroyed. So the recipe had to be altered, still keeping it very simple, yet creating an aura of labor, measurement, and finicky detail. It turned out that asking the cook to remove one tablespoon of water was precisely the right psychological fillip, and was well within the tolerance of error when using ordinary household measuring devices. The cook's guilt feelings were removed, her feeling of having been creative was bolstered, and the recipe sold millions upon millions of packages of the cake mix.

240° *The ten digits in the smallest Pythagorean triangle.*

1—The inradius
2—The indiameter, $a + b - c$
3—The short leg, a
4—The long leg, b
5—The hypotenuse, c
6—The area, $ab/2$
7—The sum of the legs, $a + b$

8—The short leg plus the hypotenuse, $a + c$
9—The long leg plus the hypotenuse, $b + c$
0—The esoteric significance of it all

—CHARLES W. TRIGG

241° *A three-part "catch" by Henry Purcell, 1731.*

When V and I together meet
We make up 6 in House or Street,

Yet I and V may meet once more
And then we 2 can make but 4,

But when that V from I am gone
Alas poor I can make but one.

[From "The Catch Club or Merry Companions, being a Choice Collection of the Most Diverting Catches for Three and Four Voices," Part 1. Reprint, Da Capo Press, N.Y., 1965.]
Originally, a catch was a round for three or more unaccompanied voices, written out as one continuous melody, each succeeding singer taking up a part in turn. Later, such a round on words combined with ludicrous effect. (From *Webster's New International Dictionary of the English Language,* second edition, unabridged, 1956.)—JOHN F. BOBALEK

242° *A curious permutation of digits.* The history of modern America began in 1492, the year the Italian navigator Christopher Columbus reached the New World. The atomic age began in 1942, the year the Italian physicist Enrico Fermi achieved the first nuclear reaction.

243° *Evaluating some desirable characteristics.* Jean Jacques Rousseau (1712–1778), the French writer and political philosopher whose ideas helped to inspire the leaders of the French Revolution, was once asked by a young lady, "What characteristics must a young lady have in order to make her man happy?" Rousseau wrote on a piece of paper:

Beauty: 0
Housekeeping ability: 0
Wealth: 0
Good nature: 1

He explained: "If a girl has nothing but good nature, she has 1. If she has other fine qualities she can be valued at 10, 100, 1000, etc. However, without the 1 in front, she is nothing."

244° *An interesting use for the new math.*   The following real estate ad appeared in *Down East* (the Magazine of Maine), April, 1973, p. 80.

> Prime Location.   Enjoy an equally lovely salt water view from almost every window all year round in this sturdily built Colonial. You will want to paint and paper, but it's priced accordingly. Charming fireplace, 1-½ baths, 4 bedrooms (or 5 if you've taken new math), garage, tool shed, low taxes.
>
> —JANET B. GOODHUE, INC.

245° *Misinterpretation.*   A man and his wife, both in their late 60s and retired, were shopping in a local supermarket. The wife stopped at the magazine rack, studied the display, selected a magazine and put it in her shopping basket. A few minutes later she returned to the rack and replaced the magazine. "Don't you want to buy that?" asked her husband.

"Not now," she replied. "I was intrigued by the title, 'Sex in the 70s.' But then it dawned on me, they mean the 1970s."

246° *An expression for forty*   One Malinke expression for "forty" is *dibi*, "a mattress," from the union of the forty digits, "since the husband and the wife lie on the same mattress and have a total of forty digits between them," to quote Delafosse.

—CLAUDIA ZASLAVSKY

247° *Rational digits again.*　In Item 1° of *Mathematical Circles Revisited,* we presented some imagined, but unhistorical, explanations of the origin of our digit symbols. Figure 4 shows still another such "rational" explanation.

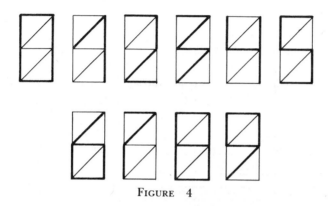

FIGURE 4

248° *In Rome.*　On the popular TV program *Hollywood Squares,* Paul Linde was asked, "If you are in Rome and a woman comes up to you and says she is XL, how old is she?" Paul Linde's instantaneous reply was, "Who cares, she's extra large."

249° *Beasting the author.*　John F. Bobalek submitted the following to English key: HOWARD W. EVES, A PROFESSOR OF MATHEMATICS AND DOCTOR OF PHILOSOPHY.

250° *Beasting Hitler.*　One can easily beast *Hitler* by adding up the numerical values of the six letters, counting A as 100, B as 101, C as 102, and so on.

251° *An odd comment.*　Upon hearing that there are no odd perfect numbers less than 10^{100}, Professor John Mairhuber commented: "That seems to indicate that odd perfect numbers are not abundant." To this, Reverend Thomas Vining remarked: "True, but if a perfect number *is* abundant, it is certainly odd."

252° *There are no irrational numbers.*   R. F. Jolly has reported the following dialogue he once had with a student:

Student: Irrational numbers do not exist!

Jolly: Oh?

Student: God is a perfect rational being; therefore all of His works are rational. Hence, irrational numbers do not exist.

253° *The ubiquitous golden ratio.*   The golden ratio, $\phi =$ 1.61803398 . . . is an irrational number that appears in many diverse situations. One of the strangest attempts to show the ever-present nature of ϕ was made by one F. A. Lonc, of New York City. He measured the heights of 65 women, and the heights of their navels. He reported that the average of the woman's height divided by the height of her navel was 1.618. It has been said that women whose measurements were not close to the golden ratio testified to hip injuries or other deformities.

LOGARITHMS

254° *A book of sensible numbers.*   Once, when the German satirist Gottlieb Wilhelm Rabener (1714–1771) called on Abraham Gottlieb Kästner (1719–1800) at Göttingen, he found a table of logarithms on Kästner's desk. Thumbing through the table he commented, "There isn't a sensible word in the whole book." "No," replied Kästner, "but there are many sensible numbers."

255° *A change in education.*   Kiano Fisher (1824–1907), the famous German philosopher and historian, has told an amusing story about logarithms. It seems that a couple of his schoolmates had a simple-minded uncle who would occasionally glance at their homework, even though he could not understand anything of what he saw. He once came upon the lads while they were using Vega's *Table of Logarithms,* a book which from cover to cover contains only numbers. His curiosity aroused, the uncle asked the boys as to the contents of the book, and they told him it contained all the house numbers in Europe. That evening, at the local beer garden, the uncle told some old friends of his that while there was a great deal to learn in school in his day, it was nothing compared to present demands—the children now had to learn all the house numbers of Europe.

256° *Graphomaths.*   There was once a popular novel in which the author created a mathematics professor who, as a student, had proved his mathematical qualifications by memorizing Vega's enormous table of logarithms, and as a result suffered a nervous breakdown.

Augustus De Morgan coined the term "graphomath" for a person who, ignorant of mathematics, attempts to describe a mathematician. Sir Walter Scott was a graphomath when he had his dreamy calculator, Dave Ramsay, swear "by the bones of the immortal Napier." Another graphomath had his mathematical hero constantly "doing sums," and still another claimed his hero was always engaged in "calculating double differentials."

257° *A sad and tragic end.*   Baron Georg von Vega (1754–1802) was a military officer and mathematician, famous for his military campaigns and his great table of logarithms. He was reportedly murdered in 1802 for "his money and his watch."

258° *An important work.*   The invention of logarithms and the calculation of the earlier tables form a very striking episode in the history of exact science, and, with the exception of the *Principia* of Newton, there is no mathematical work published in the country which has produced such important consequences, or to which so much interest attaches as to Napier's *Descriptio.*

—J. W. L. GLAISHER
Encyclopedia Britannica, 9th Edition

ARITHMETIC

259° *Cauchy and the calculating prodigy.*   The French shepherd boy and calculating prodigy, Henri Mondeux (b. 1826), gave a performance in 1840, when he was fourteen years old, before the students and faculty of the École Polytechnique. The performance started with questions from the audience. Mondeux solved the problems quickly and correctly. At length a problem was proposed that involved long calculations. Mondeux sat in deep concentration, mentally working on the problem. He seemed to be almost finished when a man in the audience stood up and triumphantly gave the answer. It was the mathematician Augustin-Louis Cauchy (1789–1857).

To protect the young prodigy, the physicist Gaspard Gustave de Coriolis (1792–1843), who was also present, challenged Cauchy to present the boy with a problem to solve. Cauchy asked the lad to find the sum of the fourth powers of the first twenty natural numbers. Mondeux closed his eyes and proceeded with the calculation. Cauchy also closed his eyes, but suddenly opened them and called out, "722,666." Cauchy, of course, had not followed the tedious path of his young rival. He used, instead, the formula

$$\sum_{i=1}^{n} i^4 = \frac{n(n + 1)(2n + 1)(3n^2 + 3n - 1)}{2 \cdot 3 \cdot 5}$$

Cauchy was able to give the desired result by computing the product of 574 and 1259, which he was able to do in his head rather easily.

—ELWOOD EDE
(freely translated from W. Ahrens, *Mathematiker-Anekdoten*)

260° *The Human Computer.* Willem Klein (b. 1912) is known as "The Human Computer." As the son of a Jewish doctor living in Amsterdam during the war, he and the rest of the family had to go into hiding from the Nazis. It was during this period of hiding that Willem Klein discovered a passion for numbers, and by much practice became highly skilled at long mental calculations. After the war he made a living performing in European music halls as "The Man with the 10,000-Pound Brain." In 1958 CERN (European Center of Nuclear Research), located near Geneva, recognized his unusual talents and hired him, for at that time he was far more efficient than the center's computer.

Still very much of a showman, Klein recently set out to break his own record of 3 minutes and 43 seconds, succeeding by one full minute. Before a rapt audience at the CERN auditorium, Klein mentally computed the 73rd root of a 499-digit number. At the end of 2 minutes and 43 seconds he announced his answer, 6,789,235. His answer was confirmed by a modern electronic computer.

Since modern computers have overtaken Klein in speed, the great mental computer intends this year (1976) to retire from CERN, return to Amsterdam, visit the schools, and "Show children how to have fun with numbers."

261° *Gender numbers.*   Adopting the Pythagorean convention that even numbers are "female" and odd numbers are "male," the following observations can be made:

1. All males are odd.
2. For any two males, there's a female (it just adds up!).
3. There's only one prime female.
4. Any female is twice a male.
5. When females start multiplying their numbers, a male can't get in edgewise.
6. Between any two females there is at least one male.
7. Any prime female has the potential to be a two-timer.
8. A male always follows a female and vice versa.
9. There is at least one prime factor in every female which is not a factor common to any male.
10. Taking all into consideration, when you get down to the roots of the matter, males and females are equally irrational.

—STEPHEN J. TURNER

262° *Oversensitivity.*   It is said that Picasso, at the age of eleven, still could not do arithmetic, because the numeral 7 looked to him like a nose upside down.

263° *A good excuse.*   When the teacher asked Johnny where his arithmetic homework paper was, the boy explained: "On my way to school I made an airplane out of it, and someone hijacked it."

264° *Why not?*   Did you hear about the plant in the math teacher's room? It grew square roots.

265° *To tell the truth.*   On his arithmetic test, Johnny worked a problem three different ways and obtained three different answers. Perplexed as to which was the correct answer, he finally in exasperation muttered: "Will the real answer please stand up?"

266° *Close, indeed!*   In my arithmetic test this morning I was mighty close to the right answers. They were only two seats away.

267° *The day's accomplishment.*   "What did you learn in arithmetic today?" asked Johnny's mother.
"I learned that seven and seven makes fifteen," Johnny replied.
"But that's not right," said his mother.
"Oh? Well then I didn't learn anything," said Johnny.

268° *Method.*   A little girl was doing her arithmetic homework on the living room floor and kept calling to her mother in the kitchen for assistance. "What is 129/3?" she called. Her mother gave her the answer. "What is 196/7?" she next called, and her mother gave her the answer. So it went for a number of problems when the mother finally asked, "Why don't you work out some of the problems yourself?" "Oh," explained the little girl, "our teacher said we could use any method we wished."

269° *Addition.*   The students were sent to the board to add 18, 27, and 34, while the teacher circulated around to see how they were doing. When she came to the overgrown athlete of the class, she found him in his gym jersey standing pleased beside his work. On the board he had figured:

$$\begin{array}{r} 18 \\ 27 \\ \underline{34} \\ \text{HIKE} \end{array}$$

270° *A difficult problem.*   H. S. Vandiver, the leading expert on Fermat's last theorem and related problems, discussed in a paper the problem of deciding whether there are infinitely many primes p for which $(p - 1)! + 1$ is divisible by p. He remarked: "This question seems to me to be of such a character that if I should come to life at any time after my death and some mathematician were to tell me that it has definitely been settled, I think I would immediately drop dead again."—LEO MOSER

QUADRANT FOUR

From Schopenhauer on arithmetic
to a matter of punctuation

COMPUTERS

MANY of the stories one hears about the sophisticated electronic computers tend to ridicule the machines one way or another. Perhaps this merely reflects a human awe and fear of the machines. At any rate, here are some more stories about computers.

271° *Schopenhauer on arithmetic.* Schopenhauer described arithmetic as the lowest activity of the spirit, as is shown by the fact that it can be performed by machine.—LEO MOSER

272° *E. T. Bell on electronic computers.* Eric Temple Bell was born in Scotland, in 1883, and received his early education in England. In 1902 he came to the United States, completed his undergraduate work at Stanford University, and then secured his A.M. and Ph.D. degrees at the University of Washington and Columbia University. After a succession of teaching posts at the University of Washington, the University of Chicago, and Harvard University, he was appointed, in 1926, Professor of Mathematics at the California Institute of Technology. He became a potent and influential force in American mathematics. He was a noted number theorist and had a deep interest in contemporary science. As a singularly fluent, gifted, highly knowledgeable, and, when he wished, skillfully cutting writer of semipopular books on mathematics and its history, he became widely known. In addition to his technical works, he also wrote, under the pseudonym of John Taine, several very successful science fiction novels. He died in 1960.

Following are a pair of remarks made by Dr. Bell about the modern electronic computers:

1. "What I shall say about these marvelous aids to the feeble human intelligence will be little indeed, for two reasons: I have always hated machinery, and the only machine I ever understood was a wheelbarrow, and that but imperfectly."

2. "I cannot see that the machines have dethroned the Queen. Mathematicians who would dispense entirely with brains possibly have no need of any."

273° *The guilty party.*   A while ago, in 1971, a newspaper reported that the employees of a large British firm became annoyed by a prolonged series of blunders emanating from the pay office, and they threatened to strike if the perpetrator of the blunders was not fired. Finally the firm gave in and fired the guilty party, a modern electronic computer, and peace and happiness were restored when human bookkeepers and accountants returned to the job.

274° *A comparison.*   Two mathematicians stood dwarfed before a monstrous computer that filled up a whole large wall. After a few seconds of running time, the machine disgorged a slip of paper. One of the mathematicians took the slip, studied it thoughtfully for a moment, turned to his companion and said, "Do you realize that it would take a corps of 500 skilled mathematicians, each writing at the rate of three digits per second and working day and night around the clock, over 1500 years to make a mistake of this size?"

275° *A cartoon by Dunagin.*   A young and pretty second-grade school teacher, standing behind her desk, addresses a little boy standing next to his front-row seat and holding a small battery-operated machine. The teacher says: "No, Rodney, you must tell me what two plus two is without referring to your electronic calculator."

276° *Translation of languages by a computer.*   There is a widely circulated story about an engineer who programmed a computer to translate from any language to any other language. To demonstrate his program at a technical gathering, the engineer asked for someone in the audience to suggest a phrase for the machine to translate, and obtained "Out of sight, out of mind." He entered this phrase into the computer and then asked for someone to suggest a language into which the phrase should be translated, and obtained "Russian." This instruction was put into the machine, and shortly the computer printed out a Russian phrase. But no one present understood Russian, and so the audience was perplexed as to the satisfactoriness of the translation. Finally someone came up with the bright idea that the Russian phrase should now be fed into the machine with instructions to translate it back into English. The engineer did this, and after a moment the machine printed out that "Out of sight, out of mind" first translated into Russian and then back into English resulted in the phrase "Blind idiot."

A similar experiment with "The spirit is willing, but the flesh is weak," resulted in "The whiskey's O.K., but the meat is lousy."

277° *The sophisticated computer.*   Visitors are welcome at a firm that makes modern sophisticated electronic computers, and twice a day in a little auditorium a free forty-minute lecture is given during which interested visitors may hear about the marvels of the machines and see one of them put through its incredible paces. Seated at a small desk at the entrance of the auditorium is an attendant who, as visitors enter the auditorium, records on a little pad, marks that look like the following:

278° *Understanding.*   At a recent exhibition of electronic computing machines in the Netherlands, Queen Juliana remarked that not only could she not understand these machines, but she could not understand the people who could understand them.

—LEO MOSER

279° *The longest root.*   A real number is called *simply normal* if all ten digits occur with equal frequency in its decimal representation, and it is called *normal* if all blocks of digits of the same length occur with equal frequency. It is believed, but not known, that π, e, and $\sqrt{2}$, for example, are normal numbers. To obtain statistical evidence of the supposed normalcy of the above numbers, their decimal expansions have been carried out to great numbers of decimal places.

In 1967, British mathematicians, working with a computer, carried the decimal expansion of $\sqrt{2}$ to 100,000 places. In 1971, Jacques Dutka, of Columbia University, found $\sqrt{2}$ to over one million places—after 47.5 hours of computer time, the electronic machine ticked off the decimal expansion of $\sqrt{2}$ to at least 1,000,082 correct places, filling 200 pages of tightly spaced computer print-out, each page containing 5000 digits. This is the longest irrational root so far computed.

280° *Computerized art.*   A number of the advanced calculators allow peripheral devices to be attached to them. Since these devices are controlled by the calculator, they may be regarded as programmable. Among the devices are plotters which, using coordinates developed by the calculator under program control, can be made to draw "curves." Although only straight line segments connecting pairs of points can be drawn, by making the line segments very short, curves can be very nicely approximated.

In 1971, the Hewlett-Packard Company, a manufacturer of these latest calculator-plotter machines, sponsored an art contest for their users. The winning entries of the Calculator Art Contest were published in the Hewlett-Packard journal *Keyboard,* which is devoted to uses of the company's calculators and peripherals. The first prize was awarded to Paul Milnarich of El Paso, Texas, for his WAVES, a striking three-dimensional-looking figure of several concentric ovals formed by a succession of sinusoidal curves obtained from the equation

$$f(x,y) = [\exp(2\sqrt{(x^2 + y^2)}/1500)][\sin 2\sqrt{(x^2 + y^2)}].$$

Second prize went to G. Winston Barber of Philadelphia, Pennsylvania, for his figure EFFIGY, a weird face-like mask, wherein the equation

$$R = A_i(B + \sin n\theta_j),$$

with various limits on A and θ, was used to produce different parts of the figure. Third prize was awarded to John A. Ashbee, of

Auburn, California, for his PLAYMATE, an attractive line drawing of a seated female nude made up of 42 fitted pieces from the graph of a fourth-degree polynomial of the form

$$y = b_0 + b_1 x + b_2 x^2 + b_3 x^3 + b_4 x^4.$$

Runners-up in the contest were: SPIROGRAM, by Lt. Ronald P. Krahe of San Antonio, Texas; UNNAMED, by N. M. Baker of Greenville, Texas; RANDOM PLOT, by W. E. Shepherd of San Diego, California; ISOMETRIC REPRESENTATION OF A THREE-DIMENSIONAL OSCILLATOR, by Fred B. Otto of Orono, Maine; and INFINITY, by Peter Zimmerman of Westport, Connecticut.

281° *Computer chess.*   In his article "The robots are coming —or are they?" in the May 1976 *Chess Life & Review,* International Chess Master David Levy points out that interest in computer chess has been steadily increasing over the past two years, with a dramatic increase in chess programs being written in the United States and Canada. There have been two computer tournaments in Europe. The first World Computer Chess Championship match was held in Stockholm in 1974 and was won by KAISSA, a program written at the Institute of Control Science in Moscow. KAISSA scored 100%, finishing ahead of four entries from the United States, three from Great Britain, and one each from Austria, Canada, Hungary, and Norway.

The best-known computer chess competition is the annual tournament sponsored by A.C.M. (the Association for Computing Machinery). In October, 1975, the sixth A.C.M. competition took place in Minneapolis and was won by the program CHESS 4.4 of Northwestern University. Earlier versions of this program had won the first four A.C.M. matches, but in the fifth competition, played in San Diego in 1974, first place was taken by the Canadian program TREEFROG.

In spite of the rapid progress being made in computer chess, Levy feels confident that no program will be able to beat him in any match prior to August, 1978, and he has offered a $2500 bet to that effect. Levy feels that it will be at least 25 years before programs will play sufficiently well to earn the FIDE International Master title.

282° *Computer stamps from The Netherlands.* Computers have invaded the art world. On April 7, 1970, The Netherlands issued a set of five postage stamps containing designs made with a computer coupled to a plotter. The computer was a CORA I, at the Technological University in Eindhoven. The attached plotter was able to draw straight line segments connecting points whose coordinates were either given or calculated by part of the program. It could also draw circles or parts of circles. The program included one for rotating the coordinate axes separately through given angles, one for translation in the direction of either axis, and one for changing the scale in the direction of either axis. The CORA I can be used in combination with larger computers for drawing graphs. Two of The Netherlands stamps were made by having a part of the computing done on the EL-x8 at the Technological University.

All five stamps are 25×36 mm, and were printed in two colors in sheets of 100 by Joh. Enschedé en Zonen in Haarlem, who are also the printers of Dutch banknotes.

The 12c stamp shows an axonometric projection of a cube, the faces of which contain a central circle surrounded by a set of concentric superellipses $(x/a)^n + (y/b)^n = 1, n > 2$.

The 15c stamp employed a perspective program to draw several cubes subdivided into smaller cubes, wherein all vertical lines are omitted.

The 20c stamp contains two perpendicular line segments rotated in opposite directions through $90°$.

The 25c stamp employed an affine program on circles, resulting in three double sets of nested ellipses.

The 45c stamp is made up of four spirals winding about a common origin.

The designs of the stamps were originated by R. D. E. Oxenaar, one of the graphic artists of the Post Office Department.

283° *Formula stamps from Nicaragua.* While on the subject of postage stamps, we might narrate the following, even though there is no connection with electronic computers.

In 1971 Nicaragua issued a series of postage stamps paying homage to the world's "ten most important mathematical formulas." Each stamp features a particular formula accompanied by an appropriate illustration, and carries on its reverse side a brief statement in Spanish concerning the importance of the formula.

The first stamp in the series is devoted to the fundamental counting formula "$1 + 1 = 2$," and pictures an ancient Egyptian grasping the concept of counting. Other early mathematical achievements honored on these stamps are the Pythagorean relation "$a^2 + b^2 = c^2$," and the Archimedean law of the lever "$w_1 d_1 = w_2 d_2$," the one so basic in geometry and other so basic in engineering. As stamps honoring later achievements, there is one devoted to John Napier's invention of logarithms and one to Sir Isaac Newton's universal law of gravitation. Among more modern formulas honored on these stamps are J. C. Maxwell's four famous equations of electricity and magnetism, Ludwig Boltzmann's gas equation, Konstantin Tsiolkovskii's rocket equation, Albert Einstein's famous mass-energy equation "$E = mc^2$," and Louis de Broglie's revolutionary matter-wave equation.

It must be pleasing to scientists and mathematicians to see these formulas so honored, for these formulas have certainly contributed far more to human development than did many of the kings and generals so often featured on postage stamps.

284° *A computer triumph.* For many years (since shortly after 1850) one of the most celebrated unsolved problems in mathematics has been the famous conjecture that four colors suffice to color any map on a plane or a sphere, where in the map no two countries sharing a common linear boundary can have the same color. An enormous amount of effort has been expended on this problem and many partial results have been obtained, but the problem itself remained refractory. Then, in the summer of 1976,

Kenneth Appel and Wolfgang Haken of the University of Illinois, established the conjecture by an immensely intricate computer-based analysis. The proof contains several hundred pages of complex detail and subsumes over 1000 hours of computer calculation. The method of proof involves an examination of 1936 reducible configurations, each requiring a search of up to half a million logical options to verify reducibility. This last phase of the work occupied six months and was finally completed in June, 1976. Final checking took the entire month of July, and the results were communicated to the *Bulletin of the American Mathematical Society* on July 26, 1976.

The Appel–Haken solution is unquestionably an astounding accomplishment, but a solution based on computerized analyses of close to 2000 cases with a total of 10 billion logical options is very far indeed from elegant mathematics. Certainly on at least an equal footing with a solution to a problem is the elegance of the solution itself. This is probably why, when the above result was personally presented by Haken to an audience of several hundred mathematicians at the University of Toronto in August, 1976, the presentation was rewarded with nothing more than a mildly polite applause.

285° *Computeritis.* Unfortunately, there is a developing feeling, not only among the general public but also among young students of mathematics, that from now on any mathematical problem will be resolved by a sufficiently sophisticated electronic machine, and that all mathematics of today is computer-oriented. Teachers of mathematics must combat this disease of *computeritis*, and should constantly point out that the machines are merely extraordinarily fast and efficient calculators, and are invaluable only in those problems of mathematics where extensive computing or enumeration can be utilized.

MNEMONICS

A MNEMONIC is any aid to the memory. Mathematicians have concocted a number of mnemonics for recalling certain important principles, formulas, and numbers. For example, in elementary trigonometry classes the students learn "<u>A</u>ll <u>s</u>tudents <u>t</u>ake <u>c</u>alculus" (or, in New York State, "<u>A</u>lbany <u>S</u>tate <u>T</u>eachers <u>C</u>ollege") to recall that in quadrant one <u>a</u>ll the trigonometric functions are positive, in quadrant two only the <u>s</u>ine and its reciprocal, in quadrant three only the <u>t</u>angent and its reciprocal, and in quadrant four only the <u>c</u>osine and its reciprocal.

We have already, in Items 41°, 119°, and 145° of *Mathematical Circles Revisited,* given examples of mnemonics in mathematics. Here are some more.

286° *A French mnemonic for pi.*   By replacing each word by the number of letters it contains, the following French poem yields π correct to 26 decimal places.

> Que j'aime à faire apprendre
> Un nombre utile aux sages
> Immortel Archimède artiste ingénieur
> Qui de ton jugement peut priser la valeur
> Pour moi ton problème
> A les pareils avantages!

The last line leads to the sequence of digits 1379; for the decimal expansion of π it should be 3279.

M. H. Greenblatt says that George Gamov once wrote an article for *Scientific American* (Oct. 1955), in which he gave the first five decimals in the expansion of π as 3.14158. Later, a reader wrote to *Scientific American,* chiding Gamov for being wrong in the fifth decimal place. Gamov, in his apology, explained that the error was due to the fact that he was an atrocious speller; he remembered π from the above French poem, but he had spelled the word "apprendre" with only one "p."

287° *Another mnemonic for pi.*

> Sir, I bear a rhyme excelling
>   In mystic force and magic spelling
> Celestial sprites elucidate
>   All my own striving can't relate.
> 3.14159 / 265358 / 979 / 323846

288° *An impossible mnemonic.*   In Item 41° of *Mathematical Circles Revisited,* four sentence mnemonics are given for recalling the decimal expansion of π to a number of decimal places. For example, if in the sentence

> May I have a large container of coffee?

one should replace each word by the number of letters it contains, one would obtain π expressed to 7 decimal places (3.1415926). The most successful mnemonic in the above reference gives π to 30 decimal places. No one has ever been able to make up a mnemonic of this kind giving π to more than 31 decimal places. Why is this?

289° *Rational mnemonics.*   The number π can be approximated by rational numbers. For example:

$$22/7 = 3.14|28,$$
$$355/113 = 3.141592|92,$$
$$104348/33215 = 3.141592653|92142,$$
$$833719/265381 = 3.14159265358|108,$$

which, in turn, give π correct to 2, 6, 9, and 11 decimal places. In the July 1939 issue of *The Mathematical Gazette,* the following mnemonics were given for recalling the last two fractions:

$$\frac{\text{calculator will get fair accuracy}}{\text{but not to } \pi \text{ exact}},$$

$$\frac{\text{dividing top lot through (a nightmare)}}{\text{by number below, you approach } \pi}$$

290° *A mnemonic in trigonometry.* Every trigonometric identity remains valid when each trigonometric function is replaced by the corresponding hyperbolic function provided we change the SIGN of each term that contains a product of *two* SINES.

—GEORGE POLYA and GORDON LATTA
Complex Variables
John Wiley & Sons, Inc., 1974, p. 69

291° *Recalling the Poisson distribution.* The Poisson distribution states that if m happenings occur on the average, then the probability that n will occur is

$$P_m(n) = (m^n e^{-m})/n!.$$

M. H. Greenblatt has noted the following remarkable mnemonic, involving, in order, the first six letters of the word *mnemonic* itself, for recalling this formula:

(m to the n, e to the minus m, over n factorial).

292° *Easy.* A mathematician had trouble remembering the local call number, 2592, of his home telephone number. Finally, after several hours of concentrated work, he announced, "I have it. The number 2592 is the unique solution in distinct positive integers x, y, z of the expression $xyzx = x^y z^x$, where the left side represents a decimal expression and the right side is a product of powers."

M. H. Greenblatt has reported this as a true story. The phone number concerned was that of B. Rothlein, who, between 1943 and 1947, lived on 41st Street in Philadelphia.

THE NUMBER THIRTEEN

293° *The one-dollar bill and the number thirteen.* How many people know the following about the United States one-dollar bill?

1. The incomplete pyramid on the back has 13 steps.

2. Above the pyramid appear the words "Annuit Coeptis," which contain 13 letters.

3. The American bald eagle holds in one talon an olive branch with 13 leaves, in the other talon a bundle of 13 arrows.

294° *Richard Wagner and the number thirteen.*   Richard Wagner's name contains 13 letters. He was born in 1813 and $1 + 8 + 1 + 3 = 13$. He composed 13 great works of music. *Tannhäuser,* one of his greatest works, was completed on April 13, 1845, and it was first performed on March 13, 1861. He finished *Parsifal* on January 13, 1842. *Die Walküre* was first performed in 1870 on June 26, and 26 is twice 13. *Lohengrin* was composed in 1848, but Wagner did not hear it played until 1861, exactly 13 years later. He died on February 13, 1883; the first and last digits of this year form 13.

295° *A prosaic explanation.*   Claude Terrail, the proprietor of the luxurious La Tour d'Argent Restaurant, overlooking the Seine in Paris, explains the superstition about having thirteen people seated at a table. "The reason is," he says, "that most people have sets of only twelve knives, forks, and dinner plates."

MERSENNE NUMBERS

296° *Father Mersenne and his numbers.*   Numbers of the form $M_n = 2^n - 1$, where $n = 1, 2, 3, \ldots$, are called Mersenne numbers, after Father Marin Mersenne (1588–1648), a Minimite friar who taught philosophy and theology at Nevers and Paris and who maintained an assiduous correspondence in an almost indecipherable handwriting with the top mathematicians of his day. Although Mersenne was a voluminous writer, it was in the theory of numbers, particularly in connection with prime and perfect numbers, that he left his most lasting mark.

Mersenne numbers are interesting for two reasons—the largest known prime numbers are Mersenne numbers, and it is with the help of prime Mersenne numbers that we discover perfect numbers (numbers which are equal to the sum of all their natural divisors that are less than the numbers themselves).

In a statement in the preface of his *Cogitata physico-mathematica*, published in 1644, Mersenne implied that the only values of n not greater than 257 for which M_n is prime are

$$2, 3, 5, 7, 13, 17, 19, 31, 67, 127, \quad \text{and} \quad 257.$$

This statement of Mersenne's has proved to be incorrect, both by omission and commission. For it has been shown that for $n = 61$, 89, and 107, M_n is prime, and for $n = 67$ and 257, M_n is composite. The complete list of currently known values of n for which M_n is prime is

$$2, 3, 5, 7, 13, 19, 31, 61, 89, 107, 127, 521, 607, 1279,$$
$$2203, 2281, 3217, 4253, 4423, 9689, 9941, 11212, \quad \text{and}$$
$$19937.$$

It took modern electronic computers to establish the primality of the very large numbers M_n for those values of n in the list exceeding 127; the last and largest one was established in 1971.

The primality of the 19-digit number M_{61}, the first case overlooked by Mersenne, was established by P. Seelhoff in 1886 and J. Pervušin in 1887. The composite nature of M_{67}, the first incorrect case included by Mersenne, was established by E. Fauquembergue in 1894 (by a process not yielding any factors of the number) and by F. N. Cole in 1903 (by expressing M_{67} as a product of two large prime numbers).

For more than 75 years, until 1952, the 39-digit number M_{127} was the largest known prime number.

297° *A well-received paper.* E. T. Bell, in his engaging book *Mathematics—Queen and Servant of the Sciences,* relates an amusing story in connection with the paper given by F. N. Cole (1861–1927) in 1903 on the factorization of the Mersenne number $M_{67} = 2^{67} - 1$.

At the October, 1903, meeting in New York of the American Mathematical Society, Cole had a paper on the program with the modest title *On the factorization of large numbers.* When the chairman called on him for his paper, Cole—who was always a man of very few words—walked to the board and, saying nothing, proceeded to chalk up the arithmetic for raising 2 to the sixty-seventh power. Then he carefully subtracted 1. Without a word he moved over to a clear space on the board and multiplied out, by longhand,

$$193{,}707{,}721 \times 761{,}838{,}257{,}287.$$

The two calculations agreed. . . . For the first and only time on record, an audience of the American Mathematical Society vigorously applauded the author of a paper delivered before it. Cole took his place without having uttered a word. Nobody asked him a question.

Some years later, in 1911, Bell asked Cole how long it had taken him to crack M_{67}. Cole replied: "Three years of Sundays."

298° *A pastime.* To while away time during the occupation in the Second World War, the French mathematician Paul Poulet (d. 1946) worked out the prime decomposition of the large Mersenne number $M_{135} = 2^{135} - 1$, finding

$$M_{135} = (7)\ (31)\ (73)\ (151)\ (271)\ (631)\ (23{,}311)\ (262{,}657)$$
$$(348{,}031)\ (49{,}971{,}617{,}830{,}801).$$

BUSINESS MATHEMATICS

299° *Obtaining a receipt.* "I loaned a tricky competitor $1000," sighed a businessman, "and he has not returned me a receipt. What can I do?"

"Write sternly," advised his friend, "and demand immediate payment of the $2000."

"You did not hear me correctly," replied the businessman. "I said the loan was $1000."

"I know," nodded the friend, "and your competitor will indignantly write you and tell you so. Then you will have your receipt."

300° *Honest business.* An angry customer returned to a jewelry store and demanded a refund on a watch that he had bought there a few days earlier. "This watch," he fumed, "loses fifteen minutes every hour."

"Of course it does," agreed the jeweler. "Didn't you see the sign '25 percent off' when you purchased it?"

301° *No accounting.* A recent newspaper ad of the Oklahoma School of Accountancy was headed: "Short course in Accountancy for women." Not long after the ad appeared, a note reached the school's president. It read: "There is no accounting for women."—*Tulsa Tribune*

302° *Mathematics of finance.* A high-school math teacher received a form letter from a loan company stating, "Because you are a teacher you can borrow $100 to $1000 from us simply by mail." The teacher's reply said, "Perhaps I can borrow from you because I am a teacher, but I would not be able to pay it back to you for the same reason."

303° *Cinderella's pumpkin.* One evening a professor from the Harvard Business School told his six-year-old son the story of Cinderella. The youngster paid close attention, particularly when the father came to the part of the story where the pumpkin turns into a golden coach. "Say, Dad," he interrupted at that point, "Did Cinderella have to report that as straight income, or was she permitted to call it a capital gain?"

304° *Depreciated currency.* When I was a kid, ten cents was a lot of money. My, how dimes have changed.

305° *A visit to a scientific shrine.* Two American physicists touring Italy visited the famous Leaning Tower of Pisa, since it was here that Galileo dramatically demonstrated that, contrary to the teachings of Aristotle, bodies of different weights dropped from the same height reach the ground in the same time. While parking their car, a uniformed attendant handed the scientists a pink ticket and collected 100 lire. When the scientists later returned to their hotel they inquired of the concierge, "Who gets the money collected for parking at the Leaning Tower?"

The concierge looked puzzled, examined the ticket, smiled,

and exclaimed, "There's no parking charge anywhere in Pisa. What you did was to insure your car against damage should the Leaning Tower fall on it."

PROBABILITY AND STATISTICS

306° *An easily understood picture.*   Suppose the nearly three billion persons on the earth were compressed into a single town of 1000 inhabitants. Then:

1. 303 persons would be white, 697 non-white.

2. 60 persons would represent the U.S.A., 940 all the others.

3. The 60 Americans would receive one-half the town's income, 940 the other half.

4. The 60 Americans would have a life expectancy of over 70 years, the other 940 a life expectancy of under 40 years.

5. The 60 Americans would consume 15 percent of the town's food supply, and the lowest income group of the Americans would be better off than the average of the 940.

6. The 60 Americans would use 12 times as much electricity, 22 times as much coal, 21 times as much oil, 50 times as much steel, and 50 times as much equipment as all 940 remaining members of the town.

307° *Hydrodamnics.*   There seems to be a newly emerging branch of probability that may be called *hydrodamnics,* concerned with the persistent cussedness of inanimate objects. It is hoped that hydrodamnics will finally resolve the reasons why:

1. a piece of buttered bread always falls to the floor with the buttered side down;

2. juice squirting from a grapefruit always goes directly into one's eyes;

3. a wind always blows out your last match when lighting a campfire.

Any reader can easily continue the list.

308° *Statisticians.* It was Mark Twain who divided all prevaricators into three classes: liars, damn liars, and statisticians. Someone else described a statistician as a person who, with his head lying in a heated oven and his feet packed in ice, says, "On the average, I feel fine."

309° *An exercise in probability.* A wise man was visited by a large delegation of malcontents who poured out their troubles to him. The wise man said, "Each of you write down your greatest trouble on a piece of paper." He then collected all the papers and dropped them into a large pot. "Now each of you draw a paper from the pot. By the laws of probability, essentially all of you will have a brand-new trouble to worry over."

The malcontents followed the wise man's suggestion. When they read the new troubles, each one begged to have his own trouble back.

310° *The odds.* A Las Vegas visitor awoke in his hotel one night with severe pains in the stomach, and put in an emergency call for the house physician. After a quick examination, that gentleman folded his stethoscope and said, "I'll give you four to one you have acute appendicitis."

311° *A safe risk.* An upset insurance inspector was scolding a new agent. "Why in the world did you write a policy on a man 98 years old?" he demanded. "Well," replied the agent, "I consulted the census report and found that there were only a very few people of that age who died each year."

312° *Statistics.* It has been said that one in every four Americans is unbalanced. Just think of your three closest friends. If they seem all right, then you're the one.

313° *A random choice.* A fellow insists his name is Seven-and-One-Eighth Flannery. He explains that his parents picked his name out of a hat.

314° *Too risky.*   "I'm truly sorry, Max," said the probability instructor, "but if I excuse you from class today, I'd have to do the same for every other married student in the class whose wife gave birth to quintuplets."

315° *A statistician protects himself.*   A certain statistician traveled considerably about the country giving popular lectures in his field of interest. His traveling, of course, had to be done by plane, a mode of journeying that caused the man great uneasiness, especially in view of the recent cases of bombs hidden aboard planes. To help alleviate his uneasiness, he calculated the probability of traveling on a plane carrying a bomb, and was pleased to find the probability very low. He then calculated the probability of traveling on a plane carrying two bombs, and was highly relieved to find this probability to be infinitesimal. So, thenceforth, in his travels about the country, the statistician always carried a bomb with him.

ALGEBRA

316° *Positive-negative.*   A usually soft-hearted father took a firm stand against one of his 17-year-old daughter's way-out demands. Sensing the finality of the "No" she was receiving, the daughter gave up instead of trying to wear her father down. "What's the use?" she sulked, "Daddy's in one of his positive-negative moods."

317° *A query.*   A little boy once asked, "Why is it that so many churches have plus signs on them?"

318° *A "proof" that "$(-)(-) = +$."*   I obtained the following elegant "proof" that "$(-)(-) = +$" from Raphael Mwajombe of Tanzania:
Imagine a town where good people are moving in and out and bad people are also moving in and out. Obviously, a good person is $+$ and a bad person is $-$. Equally obvious, moving in is $+$ and

moving out is —. Still further, it is evident that a good person moving into town is a + for the town; a good person leaving town is a —; a bad person moving into town is a —; and, finally, a bad person leaving town is a +. Our results are neatly summarized in the following table:

	moving in (+)	moving out (−)
good person (+)	+	−
bad person (−)	−	+

—ROY DUBISCH
The Mathematics Teacher, Vol. LXIV, No. 8 (Dec. 1971), p. 750

319° *Some nonassociative phrases.*

1. He is in the [high (school building)].
    He is in the [(high-school) building].

2. They went into the [dark (green house)].
    They went into the [(dark green) house].

3. They don't know how [good (meat tastes)].
    They don't know how [(good meat) tastes].

320° *Much ado about nothing.* Professor Morris Marden, who in 1975 retired from the University of Wisconsin at Milwaukee, spent over forty years studying the *zeros* of functions. As a result he has long been teased as the world's expert on *nothing.*

321° *The coconut problem.* The following is a well-known problem: "Five sailors, *A, B, C, D, E,* gathered a great pile of coconuts, which they agreed to divide equally the next morning. They had a pet monkey. Sailor *A* thought he would make sure of his share, and got up secretly in the night, divided the nuts into five equal piles, and finding an extra one left, gave it to the monkey.

Then, concealing his pile, he heaped the other four piles into one and went back to sleep. Shortly afterwards, B woke and did just what A had done: he made five equal piles, giving an odd coconut to the monkey, concealed his own pile, heaped the other four piles together, and went back to sleep. Next C awoke and did the same; then D did likewise; and finally so did E. In the morning they did as they had planned, and this time the nuts came out even in five equal shares with no extra one for the monkey. Find the (minimum) number of coconuts originally gathered."

Interest in this problem received great popular stimulation in 1926, for in that year, in the October 9th issue of *The Saturday Evening Post*, the entertaining writer Ben Ames Williams used the problem in a story. In the story the problem was employed to distract a man who should have been calculating on a contract, and it prevented the man from getting an important bid in on time.

Answer: The minimum number is 3121. For four sailors the minimum number would have been 765; for three sailors, 25; for two sailors, 11.—WILLIAM R. RANSOM

322° *A clever solution.* Enrico Fermi (1901–1954), who designed the first atomic piles and in 1942 produced the first nuclear chain reaction, has been credited with the following singularly clever solution of the coconut problem considered in Item 321°.

The number -4 is clearly a mathematical solution, because the first sailor, finding -4 coconuts, gives 1 to the monkey, leaving -5, then takes his fifth, leaving -4 again for the next sailor, and so on. To get the smallest positive solution, we must add the smallest positive number divisible by 5 five times, which is 5^5. Thus the answer to the problem is

$$5^5 - 4 = 3125 - 4 = 3121.$$

323° *A coach's formula.* Mr. Webb, a very successful coach at St. John's, Cambridge, was a good teacher and an amusing personality. He used to say, "*pp > jj*" (plodding patience is greater than jumping genius). He could be very sarcastic, and would say to a student, "Write all you know on this piece of paper"—something about an inch square.—L. J. MORDELL

324° *The metric system.*   Part of the work in present-day algebra classes is converting from English measurement to metric measurement, and vice versa.

One of Dunagin's masterful cartoons (it appeared in the Friday, September 12, 1975 issue of the *Bangor Daily News*) shows a man enthusiastically addressing members of the U.S. Government Dept. of Commerce. With charts and conversion plans at hand, he is stressing the importance of the coming change-over from the English to the metric system of measurements. With a warning finger in the air, he says: "I want everyone to think metric! Remember an ounce of prevention is worth a pound of cure!"

325° *Pointing the way.*   The nation's first metric-system highway signs were erected along Interstate 71 in Ohio. Thus a sign outside Columbus now reads: CLEVELAND—94 MILES—151 KILOMETERS. The new signs are intended to prepare motorists for the nationwide conversion, which is expected to occur by 1983.
—Tomorrow's World

326° *Andre Weil's exponential law of academic decay.*   "First-rate people appoint first-rate people; second-rate people appoint fourth-rate people, who appoint eighth-rate people, and so on."

GEOMETRY

327° *The fourth dimension in landscape planning.*   In designing a landscape plan for a piece of property, the width and depth of the property, the position of the buildings on the property, the rise and fall of the land, and the height of the buildings, of neighboring structures, and of the existing large trees, collectively constitute three of the dimensions that must be included in the plan. But there is a fourth dimension, the passage of time, that must also be included, and this dimension is one of the most important considerations in the design, and is a dimension that does not show up in the initial planning process.

Many a landscape design has been ruined by a failure to take care of the fourth dimension. Trees and shrubs, that in time grow very large, are planted too close together or too close to view-commanding windows. Shrubs that reach only a moderate height are later hidden by other taller-growing items. It takes real skill and vision on the part of the landscape architect to prepare a design that, when executed, will look attractive, not only on completion of grading, growing of grass, and planting of trees and shrubs from the nursery, but will continue so as time marches on. Indeed, the truly good plan will be attractive at the start and then actually gain in attractiveness with the passage of the fourth dimension. The landscape architect must constantly ask himself how his developing design will look five years hence, ten years hence, twenty years hence.

328° *Applicable to angle trisectors and circle squarers.*   William G. McAdoo, Secretary of the Treasury in President Woodrow Wilson's Cabinet from 1913 to 1918, claimed: "It is impossible to defeat an ignorant man in argument," a statement certainly applicable to angle trisectors and circle squarers. Apropos these latter people, Garrett Hardin, in his excellent book *Nature and Man's Fate,* says: "Those who have had any contact with angle trisectors, circle squarers, or the inventors of perpetual-motion machines will attest to their being a very queer bunch of people, indeed. It is not their proposals that merit study, but their personalities. What defect in their character is it that makes them unwilling to accept the idea that perhaps they cannot have everything they want? Whatever it is, it is akin to the immaturity of the spoiled child and the compulsive gambler."

329° *A strong theorem.*   In a paper on the foundations of Euclidean plane geometry, R. H. Bruck explains:
"The axioms of incidence require so little of the Euclidean plane that very few theorems have been proved. Indeed, the main

theorem—I call it Hall's theorem—might be stated as follows: Any damn thing can happen."—LEO MOSER

330° *To meet or not to meet.*   There are geometries in which parallel lines do not exist. Of such a study one may say: "The reason parallel lines never meet in this geometry is because they were never introduced."

331° *Careful use of terms.*   A geometry professor was induced to introduce the daughter of one of his colleagues to his son. To prepare the son, the professor first described the young lady to the lad. After it all was over, the boy, who hadn't been too impressed with the girl, said to his father, "I thought you said the young lady's legs were without equal." "Not at all," replied the professor, "I said they were without parallel."

332° *Victor Borge's metric.*   "Laughter is the shortest distance between two people."

333° *Martian geometry.*   Harry Schor, of Far Rockaway High School on Long Island, says that he is teaching geometry to a class of such peculiar students that he has to prove right triangles congruent by "hypotenuse and tentacle."—ALAN WAYNE

334° *Volume.*   Mrs. Jones was busily applying light yellow paint to the dark brown wall-board inside her garage. Neighbor Smith commented that the lighter color would make the garage seem larger. "I certainly hope so," said Mrs. Jones, "We could really use the extra space."

335° *Circles.*   A first-grade teacher asked her pupils to draw pictures showing what their fathers did for a living. She noticed one child drawing circles on his paper, and asked, "What does your father do for a living?"
The child replied, "He's a doctor, Mrs. Sawyer. He makes rounds."

336° *A possible misunderstanding.*   In the "continuous geometry" developed by von Neumann, points, as such, play no role. Originally von Neumann proposed another name for his geometry, but his colleagues pointed out that this name might lead to misunderstanding. Von Neumann's proposed name was "pointless geometry."—LEO MOSER

337° *An oblate spheroid.*   There was a politician who can be said to have resembled the earth. The politician, you see, was severely clobbered in a bid for re-election to Congress, and the earth is an oblate spheroid—that is, it is flattened at the poles.

338° *Angles.*   An Eskimo father sat in his igloo reading nursery rhymes to his little boy. "Little Jack Horner sat in a corner," began the father, when the boy interrupted with, "Daddy, what is a corner?"

RECREATIONAL MATTERS

339° *Loculus Archimedes.*   One of the earliest examples of a "cut-out" puzzle that has come down to us is that which has become known as the *loculus Archimedes,* or the *box of Archimedes.* It consists of assembling the fourteen pieces (11 triangles, 2 quadrilaterals, 1 pentagon) pictured in Figure 5, top, into a square. It is a teasing puzzle that makes a nice gift when the pieces are neatly cut from thin wood or plastic and are to be assembled to fill the bottom of an accompanying shallow square box, as in Figure 5, bottom.

Historians feel that we cannot ascribe the puzzle to Archimedes, and explain the attachment of his name to it as probably only a way of asserting its cleverness. Indeed, the phrase "Archimedean problem" has come proverbially to mean anything that is difficult. Since we are ignorant of the origin of the puzzle, we do not know why this particular dissection of a square was chosen.

340° *Storage place for imaginary numbers.*   During the 1976 school year, I received, through the mail, a little square box of

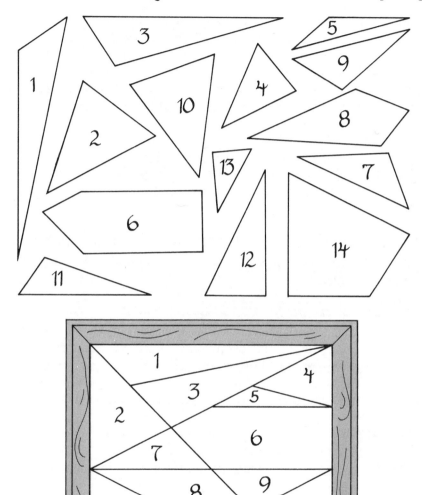

FIGURE 5

dimensions 4 inches by 4 inches by 2¼ inches deep. The box was beautifully finished and bore a small brass clasp and a pair of brass hinges. Upon opening the clasp it was found that the box had no "insides;" it was merely two pieces of wood, one ¾ of an inch thick and the other 1½ inches thick, hinged together. Lying between the top and bottom pieces of wood was a small piece of paper bearing the following rhyme:

> For a place to store imaginary numbers,
> There are very few devices.
> Although this is not a complex box,
> I feel that it suffices.

The box was a humorous gift from my former fine student, then excellent assistant, now masterful teacher, and always good friend Carroll F. Merrill.

341° *A triangular Christmas card.*   When Charles W. Trigg was serving as Dean of Instruction at Los Angeles City College, he designed and mailed to his friends and to friends of his college an attractive geometrically motivated Christmas card, pictured in Figure 6. With the three circular segments folded over and properly tucked in, the card assumes the triangular form pictured in Figure 7.

342° *An unusual Christmas card.*   Several years ago, the Reverend George W. Walker of Buffalo, New York, composed the following poem, which is circular and endless in the sense that after reading the first six lines, you are to continue by repeating these six lines over and over endlessly.

True love and friendship come as gifts of God, and we can say
The precious gifts of God endure forever, like the way
You start to read this anywhere and presently you find
A ring like this will never end, which ought to bring to mind
We start afresh on New Year's Day, but, after that, we see
The years go on eternally, and so, it seems to me,

. . . .

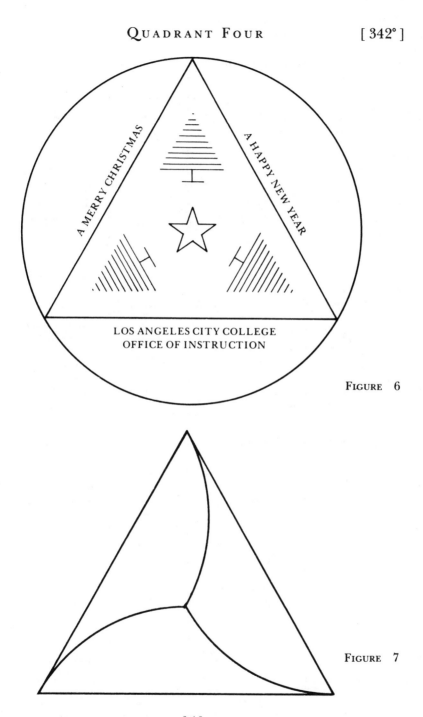

A MERRY CHRISTMAS

A HAPPY NEW YEAR

LOS ANGELES CITY COLLEGE
OFFICE OF INSTRUCTION

Figure 6

Figure 7

The Reverend now took a long strip of paper (see Figure 8) and typed the first three lines, tandem, along the middle of the strip. Next, turning the strip over (see Figure 9) he typed the next three lines, tandem, along the middle of the reverse side of the strip. Then he took the strip, gave it a half-twist, and taped the two ends together, forming a Möbius strip containing the endless poem along its midline. He now had an unusual Christmas, or New Year's, card, to send to his friends, for the Möbius strip can be neatly folded and placed in an envelope (see Figure 10).

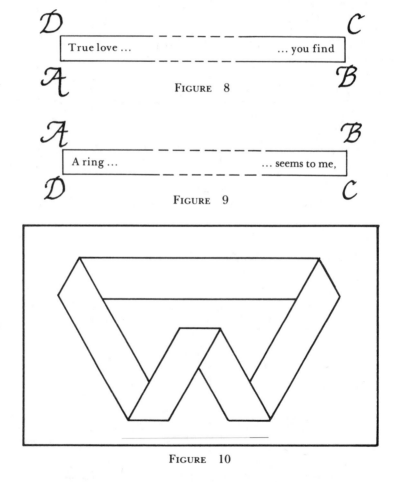

D C

True love ...   ... you find

A FIGURE 8 B

A B

A ring ...   ... seems to me,

D FIGURE 9 C

FIGURE 10

142

A Möbius strip of this sort can be made containing any circular and endless poem or piece of prose, such as the familiar circular and endless story:

The night was dark and stormy, and we were all seated around the campfire smoking, when someone said, "Captain, tell us a story." So the Captain began:

343° *Another mathematical Christmas card.*   Some years ago, Professor Richard V. Andree (of the University of Oklahoma, in Norman, Oklahoma) and his wife Josephine P. Andree mailed to their friends what has proved to be perhaps the best of the Christmas cards based upon the idea of graphing some equation or equations. Figure 11 shows the appearance of the front face of the card. The design on the card was actually in two colors—red for the coordinate axes and blue for everything else.

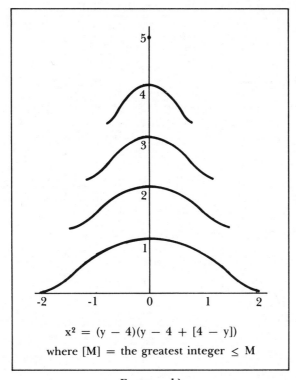

$$x^2 = (y - 4)(y - 4 + [4 - y])$$

where [M] = the greatest integer ≤ M

FIGURE   11

143

344° *Some season's greetings for the mathematics student.*
Many mathematical expressions have been concocted which, when
expanded or operated on in some way, yield familiar season's
greetings. For example, Professor Richard V. Andree designed the
following:

$$\begin{vmatrix} M & R^2 \\ -Y & \ln^{-1}1 \end{vmatrix} + XW,$$

where it must be realized that, using customary symbols of physics,

$$W = F \bullet S \quad \text{and} \quad F = MA.$$

Many years ago, when I was teaching in West Virginia, one of
my calculus students, then Miss Bertha Weaver, devised the following:

Given

$$y = (x + M/2)^2 - (E - x)^2/2 + (9y - 6)/3 + 2Rx + \int y\, dx$$
$$+ (14 - 6y)/2 + x^2M/2 - (2A - x)^2/4 + x(S - x/4).$$

Find dy/dx by implicit differentiation.

345° *A holiday greeting.* Professor Charles W. Trigg,
when at Los Angeles City College, proposed the following problem (*The American Mathematical Monthly,* Problem E 1241, Dec.
1956):

> The holiday greeting, *MERRY XMAS TO ALL,* is a cryptarithm in
> which each of the letters is the unique representation of a digit, and
> each word is a square integer. Find all solutions.

It turns out that there are only two solutions,

27556 3249 81 400   and   34225 7396 81 900.

Azriel Rosenfeld observed that if the further requirement is imposed that the *sum of the digits* of each word be a perfect square,
then the solution is unique. Edgar Karst concocted the allied problem: The holiday greeting, *MERRY + XMAS = TOALL,* is a
cryptarithm in which each of the letters is the unique representa-

tion of a digit and each word is divisible by 3. There is now a unique solution, $84771 + 5862 = 90633$.

346° *Anagrams.*   In the December 1952 issue of *The American Mathematical Monthly* appeared the "fun" problem E 1041:

> Surely the days of anagrams are not dead. Here are four well-known mathematicians: (1) A DONUT SHIP,   (2) SHE IS A NUT,   (3) SEWER STAIRS,   (4) FIRE ON SUB.

The answers, which appeared in the June–July 1953 issue of the journal, are: (1) DIOPHANTUS, (2) STEINHAUS, (3) WEIERSTRASS, (4) FROBENIUS. A good deal of ingenuity was expressed by various solvers of the problem. For example, Charles W. Trigg produced, among others, the following alternative anagrams for the above four mathematicians: (1) THUDS PIANO, I PUT SHAD ON; (2) USES A HINT, TIN HAS USE, SHUNS A TIE, THE USA SIN; (3) WATERS RISE, SWEARS RITES; (4) I BURN FOES, FIE ON BURS, O FINE RUBS.

Trigg, Paul L. Chessin, and H. W. Gould added anagrams of further mathematicians, such as: WENT ON (NEWTON), ROTTED HUN (TODHUNTER), REC'D ROE (RECORDE), IV TENS (STEVIN), KIND DEED (DEDEKIND), SEE GUARDS (DESARGUES), PELT ONCE (PONCELET), LIE CRUEL APE (PEAUCELLIER), BAN ON RICH (BRIANCHON), A FREE BUCH (FEUERBACH), RICH CASE (SACCHERI), RAM NINE (RIEMANN), SING A TUNE, MOOR (REGIOMONTANUS), I REMOVED (DE MOIVRE), MAR CONE (CREMONA), DEER GLEN (LEGENDRE), CALL APE (LAPLACE), I RIVAL ACE (CAVALIERI), SET CEDARS (DESCARTES), TIRE CHILD (DIRICHLET), HARM A DAD (HADAMARD), I CHARM ALEC (CHARMICHAEL), A CURT NO (COURANT), LEAN MUSE (MENELAUS), REACH DIMES (ARCHIMEDES), ONE LIME (LEMOINE), NO GEM (MONGE), OR AN EPIC (POINCARE), AS SHE (HASSE), HERE TIM (HERMITE), REGAL NAG (LAGRANGE), NICE CORPUS (COPERNICUS), LAMED BERT (D'ALEMBERT), A CRITIC (RICCATI), ON ETHER (NOETHER), A PINER (NAPIER), RECIPE (PEIRCE), ABLE

(ABEL), BRED IN US (BURNSIDE), I LAG SO (GALOIS), READ CHIMES (ARCHIMEDES), AM SURE NOT GO IN (REGI-OMONTANUS), CABIN FOCI (FIBONACCI), RUN A CLAIM (MACLAURIN), and several others.

On July 20, 1953, Henriette von Boeckmann, Secretary of The National Puzzlers' League, Inc. (a group that was organized on July 4, 1883), sent the following letter to the Editor of the Problem Department of *The American Mathematical Monthly:*

Dear Dr. Eves:

The National Puzzlers' League met in convention in the Hotel Statler, N.Y. City, July 3–4–5, 1953. At the business meeting one of the members, who is a subscriber to *The American Mathematical Monthly,* brought to the attention of the meeting, that your June–July issue gave some space, in their Problem Department, to a puzzle form—the Anagram.

The names of four famous mathematicians were "anagrammed," but, according to the standards of the N.P.L., they are merely muta-tions. It is true that the dictionaries define an Anagram as a rear-rangement in the letters of a world or phrase to produce some other word or phrase, but the N.P.L. narrows that meaning so that the anagrammed word must bear a relation in meaning to the original. Here are a few specimens of Anagrams as sponsored by the N.P.L.:

One smart hat—Tam O'Shanter,
So I sit and sip—Dissipations,
It's slam bang—Lambastings,
Is "anti" in "spare child"—The disciplinarians.

The N.P.L. also recognizes the Antigram, in which words express opposite meanings.

You can see that anagramming a proper name is exceedingly difficult; the result is usually a mutation.

This letter is in no sense a criticism; it is merely an attempt to keep the mental sport of puzzling on a high plane. The convention, through its President, requested its Secretary to explain to you the N.P.L.'s conception of the Anagram, which by many is considered the highest form of the puzzling art.

Yours very truly,
Henriette von Boeckmann, Secretary N.P.L.

From the point of view of The National Puzzlers' League, perhaps only WENT ON (NEWTON), KIND DEED (DEDEKIND), and ABLE (ABEL) can rate as *true* anagrams, and NO GEM

(MONGE) and I LAG SO (GALOIS) might be considered as anti-grams. It is interesting to observe that the first of our *true* anagrams was noted by Augustus De Morgan (see Item 198° of *In Mathematical Circles*).

Finally, Trigg gave what may be the only palindromic anagram of a mathematician's name: AVEC (CEVA).

347° *The great "show-me" craze.* In the mid-1970s the great "show-me" craze erupted, like an epidemic, among the mathematically literate of the United States. The disease gained rapidly and engulfed an enormous number of the mathematical community. It raged rampantly for some time, and then subsided almost as quickly as it had arisen. During its height, its taint was to be found in a number of the country's mathematics journals. Among the victims most deeply smitten was Charles W. Trigg, as was to be expected, for the disease seemed to ferret out the keenest and most agile minds of the profession. While under the frenetic fever of the derangement, Trigg produced a prodigious number of "show-me's," of which the following can be considered only as a mere sample.

Show me a useless exponent and I'll show you one.

Show me some protesters with locked arms on a hotel floor and I'll show you a connected set.

Show me a man who counts on his fingers and I'll show you a digital computer.

Show me how to partition six and I'll show you how to do it to two, too.

Show me a glib salesman and I'll show you a line.

Show me an erratic teenager with no personal income and I'll show you a dependent variable.

Show me a community with approximately equal populations of Nordics, Africans, Chinese, and Indians and I'll show you a four-color problem.

Show me an upended maple tree and I'll show you some complex roots.

Show me a singing birthday party and I'll show you a harmonic function.

Show me a youngster who has lost a trio of front teeth and I'll show you a three-space.

Another afflicted member of the mathematical fraternity was Fred Pence, who produced the following "show-me's."

Show me two cars enmeshed after an auto accident and I'll show you a rectangle (wrecked tangle).

Show me Joe Frazier and Muhammad Ali and I'll show you ex-opponents.

Show me the squares of 6, 5, and 6 and I'll show you some interesting figures.

Show me a mermaid dressed in seaweed and I'll show you an algae bra.

Other similar viruses have at times spread through the mathematical world—who can forget, for example, the "Tom Swiftie" craze, traces of which are still to be found in the mathematical hinterlands. Indeed, an enterprising individual could easily assemble an interesting booklet of case histories of these mathematical pestilences. There is no telling if some of these will, like the Asian flu, reappear in the mathematical community, nor can we guess what new virulent variations may afflict us in the future.

348° *Benjamin Franklin's semimagic square.* In *Mathematical Circles Squared* we devoted a section to magic squares; we now consider a few more of these delightful objects, and refer the uninitiated reader to Item 30° of the above reference for the meanings of commonly used technical terms.

Benjamin Franklin (1706–1790) interested himself in the construction of semimagic squares possessing the "bent diagonal" property (see Item 317° of *In Mathematical Circles*). Figure 12A shows an 8 × 8 Franklin semimagic square—each column and each row sums to 260 (unfortunately, the two principal diagonals fail to do so, and thus the square is only semimagic rather than magic). But the five bent diagonals shown in Figure 12B also each sum to 260, and there are five such bent diagonals based on the top, five based on the right side, and five based on the left side of the square, each summing to 260. In addition, the eight squares

52	61	4	13	20	29	36	45
14	3	62	51	46	35	30	19
53	60	5	12	21	28	37	44
11	6	59	54	43	38	27	22
55	58	7	10	23	26	39	42
9	8	57	56	41	40	25	24
50	63	2	15	18	31	34	47
16	1	64	49	48	33	32	17

A

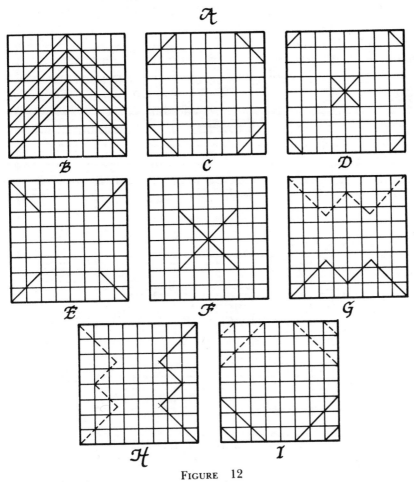

FIGURE 12

149

indicated in Figure 12C, the eight indicated in Figure 12D, the eight indicated in Figure 12E, and the eight indicated in Figure 12F also each sum to 260. The eight squares indicated by the full line of Figure 12G, and those indicated by the dashed line in Figure 12G, also each sum to 260, and a similar remark can be made of the squares indicated by the full and dashed lines of Figure 12H and Figure 12I. We thus have a total of 46 sums of 260 each. Finally, each quarter of the given square is a 4 × 4 semimagic square of magic sum 130.

349° *Dr. Frierson's pandiagonal Franklin magic square.* A magic or a semimagic square each of whose bent diagonals has the magic sum of the square is known as a *Franklin square.* The Franklin square of Item 348° is semimagic, and for some years no one produced a *magic* Franklin square. Over a century after Franklin's introduction of his squares, a Dr. Frierson constructed the 8 × 8 Franklin square shown in Figure 13, and this square is *pandiagonally magic* (that is, not only each row and each column, but also each principal diagonal and each broken diagonal, sums to 260). Dr. Fierson's square also possesses the properties of Figures 12B, 12C, 12D, 12E, 12F, and 12G. Moreover, each quarter of the square is a 4 × 4 pandiagonally magic square.

64	57	4	5	56	49	12	13
3	6	63	58	11	14	55	50
61	60	1	8	53	52	9	16
2	7	62	59	10	15	54	51
48	41	20	21	40	33	28	29
19	22	47	42	27	30	39	34
45	44	17	24	37	36	25	32
18	23	46	43	26	31	38	35

FIGURE   13

350° *A nesting magic square.* Figure 14 shows a magic square of order 7 containing concentrically within it magic squares of orders 5 and 3.

4	9	8	47	48	49	10
38	19	20	17	34	35	12
39	37	26	27	22	13	11
43	36	21	25	29	14	7
6	18	28	23	24	32	44
5	15	30	33	16	31	45
40	41	42	3	2	1	46

FIGURE 14

351° *The IXOHOXI.* The square shown in Figure 15 is a magic square with magic sum 19998 whether the square is viewed rightside up, upside down, or in a mirror. This square, for apparent reasons, is called the IXOHOXI (pronounced iks-ō-hox-ē).

8818	1111	8188	1881
8181	1888	8811	1118
1811	8118	1181	8888
1188	8881	1818	8111

FIGURE 15

A CONCLUDING MISCELLANY

352° *Houseboat.* In Item 270° of *Mathematical Circles Revisited* appears the anecdote about Sidney Cabin and the integration of *d(Cabin)/Cabin,* in which the answer comes out as "log *Cabin*" if the constant of integration is neglected. It is interesting that if the constant of integration is *not* neglected, the answer becomes "*houseboat*"—log *(Cabin)* + *C.*—MICHAEL R. VITALE

353° *Sheldon and infinity.*   C. Sheldon, of the University of Alberta, was very fond of talking about infinity. He pretended to be always on the lookout for it. Toward the end of a lecture he would draw a horizontal line on the blackboard, extend it to the end of the board and beyond, on the wall, toward the door; then he would open the door and continue the line around it and go off to his office. Next day he would return to class with cupped hands and exclaim, "I have it! I have it right here! I finally caught infinity!" Then he would stand near the open window and slowly open his hands. "Oh! Oh!" he would exclaim, "There it goes out of the window. It's gone! It's gone!"

On one occasion at this point, a student who had failed the course the previous year and to whom Sheldon's performance came as no surprise, jumped up, drew a toy pistol from his pocket, aimed the gun out of the window, and fired.

When a new library was being built at the University of Alberta, the excavators had made quite a large mound of earth. Some students mounted a sign on top of the mound which read: MT. SHELDON. YOU CAN SEE INFINITY FROM HERE.

—LEO MOSER

354° *The true founders of the Institute at Princeton.*   Carl Kaysen, a Harvard political economist who succeeded J. Robert Oppenheimer as director of the Institute for Advanced Studies at Princeton, once remarked: "It is not noted in the official history, but the true founders of the Institute for Advanced Studies were Abraham Flexner and Adolph Hitler."

355° *A suggestion.*   Mathematics instructor to an uneasy student: "If you'd rather your parents didn't know your grade, leave a space between your first and last name, and I'll fill in your letter grade as your middle initial."

356° *A lonely profession.* A botanist can easily talk interestingly about his work to a person of another field. The same can be said of a geologist, a physicist, a biologist, an historian, an artist, a musician, an economist, a philosopher, a writer, an astronomer, and so on. But it is pretty difficult for a mathematician to talk mathematics to anyone else but another mathematician. This makes mathematics a lonely profession, and accounts for why mathematicians will travel great distances to attend meetings at which other mathematicians will be present. And the deeper one is in his mathematics, the more difficult it becomes to find someone to talk to. A mathematician at a university, for example, can even be lonely among his colleagues at the university if he happens to be, say, the only mathematical logician there. They tell of a topologist who traveled 2500 miles to meet another topologist so that he might tell of some of the things he had been doing in topology. When he met the other topologist he started right in, telling about his work. The second topologist seemed not to understand anything the first one was saying, and finally blurted out: "Oh, I see, you are a *combinatorial* topologist; I'm a *differential* topologist, you see." And neither one could talk shop to the other. How often a mathematician, discovering an especially attractive proof of a theorem, longs for, but cannot find, someone to whom he can describe his solution.

357° *Overheard at a university.* (a) "My wife doesn't understand me. I'm an algebraic topologist."

(b) Commented a frustrated math professor: "He's the first student the headshrinkers ever failed with."

(c) "I guess I've lost another pupil," sighed the math professor as his glass eye slid down the drain.

358° *Headline.* The headline of a newspaper item read: UNEMPLOYED MATHEMATICS TEACHER LACKS CLASS.

359° *Point of view.* To illustrate how opinions may vary, depending upon one's point of view, consider the following. A young Roman student Cassius handed to his teacher Brutus an arithmetic problem that looked like this:

$$\times | + | = \times$$

which, from the teacher's point of view was wrong. But turn the page upside down and you'll see that from the student's point of view the problem was correct.

360° *A matter of punctuation.*   In a now forgotten novel, written in the early part of the present century, occurs the following schoolroom incident:

> "How many one-quarter-inch squares are there in a rectangle of width one inch and length one and a quarter inches?" asked Mr. Fuzzleton of his class. Immediately up shot Scott's arm, and, unable to repress himself, the lad excitedly shouted, "Twenty!"

It can be shown that 20! one-quarter-inch squares are more than enough to cover the entire United States.

L'ENVOI

*From Ebbinghaus' illusion
to equivocal figures*

GEOMETRICAL ILLUSIONS

A GOOD geometrical illusion is clever and amusing, and may be regarded as a pictorial mathematical joke or story. Indeed, a good geometrical illusion, being a condensed pictorial account of an interesting geometrical incident, can be regarded as a special kind of mathematical anecdote. It may, then, be quite proper to include some geometrical illusions in our collection of mathematical stories and anecdotes.

On the principle that it is poor showmanship to explain a joke or a story, we make little effort here to explain the geometrical illusions—we merely present them. For an analysis of why our eyes on occasion deceive us, the reader may consult M. Luckiesh's excellent work, *Optical Illusions,* reprinted by Dover Publications in 1965.

My own fascination in, and wide collection of, geometrical illusions started years ago with Luckiesh's interesting book. Just as it is my hope some day to assemble into a book my collection of mathematically motivated art (a peek at which was offered in *Mathematical Circles Squared*), it is also my hope some day to assemble into a book my collection of geometrical illusions. Here are just a few of them—all simple ones and all involving *circles.*

In looking at geometrical illusions, it must be kept in mind that what may greatly deceive one person's eyes may not so greatly, even if at all, deceive another person's eyes.

In our four trips around the mathematical circle (*In Mathematical Circles, Mathematical Circles Revisited, Mathematical Circles Squared,* and *Mathematical Circles Adieu*) we have related a total of 4(360) = 1440 mathematical stories and anecdotes, plus 9 more in an Addenda to *Mathematical Circles Squared.* So we now begin this final little collection with 1450°.

AREAL DECEPTIONS

1450° *Ebbinghaus' illusion.*   This illusion is a striking example of an areal deception. In Figure 16, though the central circles in the two parts of the figure are equal, the one surrounded by small circles appears greater than the one surrounded by large circles.

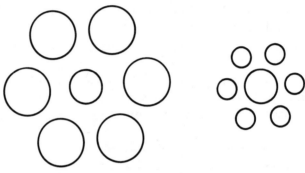

FIGURE   16

1451° *A general principle.*   A generalization of the Ebbinghaus illusion states that, of two equal figures, one adjacent to small extents and one adjacent to large extents, the former figure appears to be larger. Thus, in Figure 17, though the two interior circles are equal, the one on the left appears to be larger than the one on the right. Again, in Figure 18, though the two circles are equal, the one on the left appears larger than the one on the right.

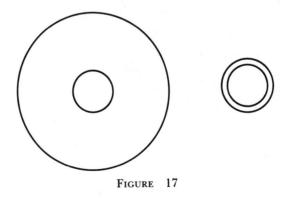

FIGURE   17

158

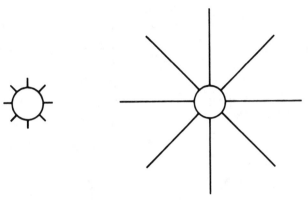

FIGURE   18

1452° *The rising moon.*   Probably everyone has observed, at one time or another, that the sun or full moon when near the horizon appears much larger than when at a higher altitude. That this is only an illusion becomes quite evident if one should view the sun or full moon, when near the horizon, through a cylindrical tube of small diameter; the sun or moon seems instantly to shrink to its size when viewed overhead. Many explanations have been offered to account for the above illusion, one being based upon Figure 19, wherein the circle close to the vertex of the angle appears larger than the other circle, though the two circles are actually equal.

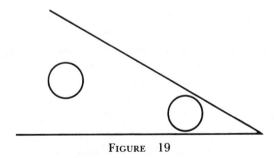

FIGURE   19

1453° *Interior and exterior extents.*   Generally, of two equal convex figures, say, one containing an adjacent extent in its inte-

159

rior and one containing an adjacent extent in its exterior, the latter convex figure will appear to be larger. This is illustrated in Figure 20, in which the interior circle on the right appears larger than the exterior circle on the left, though the two circles are actually equal. The illusion is seen even more strikingly in Figure 21, in which, though the two circles are equal, the one on the left appears larger than the one on the right.

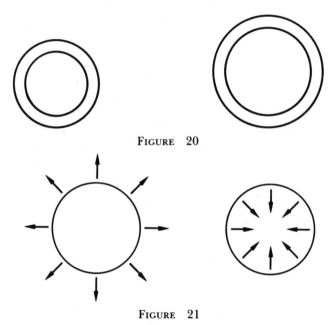

FIGURE   20

FIGURE   21

1454° *A property of the circle.* If one should draw, on the same sheet of paper, a circle and the first few regular polygons, making them all of the same *area,* the equilateral triangle would *appear* to have the greatest areal extent and the circle the least. This is seen in Figure 22. Perhaps one reason for this illusion lies in the fact that in our set of equiareal figures, the equilateral triangle has the greatest perimeter and the circle the least.

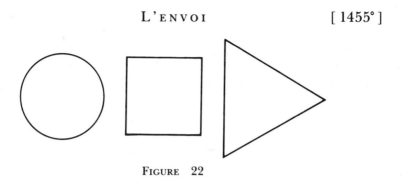

FIGURE　22

1455° *False judgments.*　A judgment of comparison between two areas can sometimes be influenced merely by how the two areas are situated with respect to one another. Thus, in Figure 23, though the two figures are congruent, the lower one appears to be somewhat larger than the upper one. On the other hand, of the two congruent figures in Figure 24, it is now the upper one that appears somewhat larger than the lower one.

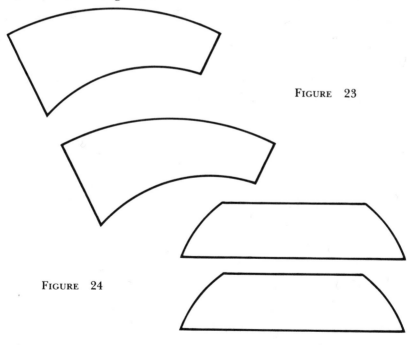

FIGURE　23

FIGURE　24

1456° *Perspective.* There are striking illusions involving perspective, or at least the influence of converging lines. An example appears in the Frontispiece of this volume, wherein the two "balls" are really equal to one another, but do not appear so to the normal eye.

LINEAR DECEPTIONS

1457° *Three circles.* In Figure 25, the distance between the inside edges of the two circles on the left is actually equal to the distance between the outside edges of the two circles on the right.

FIGURE   25

1458° *Three circles again.* To most human eyes, a horizontal line segment drawn on a piece of paper appears shorter than an equal vertical line segment drawn on the paper. Utilizing this fact, the distance illusion of Item 1457° can be strengthened as in Figure 26, wherein the distance between the inside edges

FIGURE   26

of the vertically aligned circles is actually equal to the distance between the outside edges of the two horizontally aligned circles.

1459° *Baldwin's illusion.*   In Figure 27, the little fulcrum is located at the center of the line segment joining the two circles. Though this illusion is not as positive as many others, most viewers would feel that the fulcrum is placed closer to the larger circle.

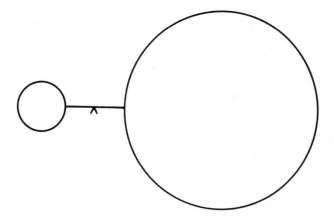

FIGURE   27

DEFORMATIONS

1460° *An interrupted circumference.*   Most people would feel that in Figure 28, the outer arc on the left of the vertical interruption is the continuation of the rest of the circle, whereas it is actually the inner arc that forms the continuation.

163

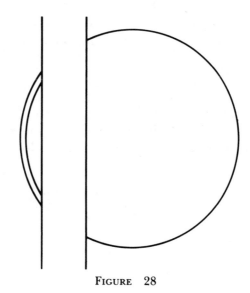

FIGURE   28

1461° *Semicircles.*   Of two equal semicircular arcs, one possessing its diameter and the other not (see Figure 29), the former will generally appear as smaller and flatter than the other.

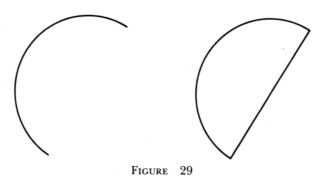

FIGURE   29

1462° *Another interrupted circumference.*   In Figure 30, the short arc of the interrupted circumference of a circle appears flatter and of greater radius of curvature than the larger arcs.

164

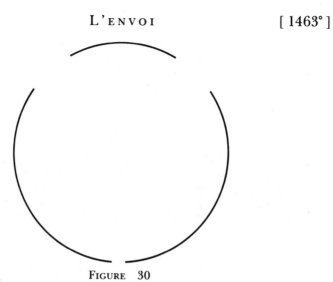

FIGURE 30

1463° *Pinched-in circumference.* To most eyes, a polygon inscribed in a circle appears to distort, or pinch in, the circle at the vertices of the polygon. This is illustrated in Figure 31, where the circumscribed circle appears to dent inward at the vertices of the hexagon. Similarly, the sides of the hexagon appear to sag inward at the points of contact with the inscribed circle.

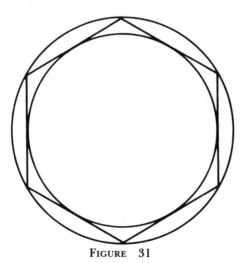

FIGURE 31

165

1464° *The sagging line.*   The circular arcs above the horizontal straight line segment of Figure 32 cause the segment to appear to sag slightly in the middle.

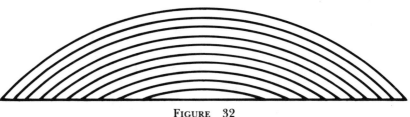

FIGURE   32

1465° *The arching line.*   By placing the circular arcs of Figure 32 below the line segment, as in Figure 33, the horizontal straight line segment now appears to arch slightly in the middle.

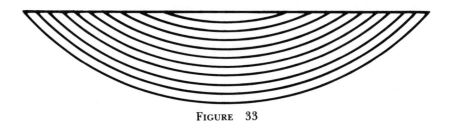

FIGURE   33

1466° *Is it really round?*   Figure 34 strikingly illustrates how a figure can appear quite deformed by merely placing it in an appropriate field. The closed curve of the figure is actually a true circle.

1467° *Equivocal figures.*   There are many geometrical illusions involving what may be called *equivocal figures*; that is, figures that can change in appearance because of the viewer's fluctuation in attention or in association. Thus most of us at one time or another have viewed the famous Schröder reversing staircase, the Thiéry reversible chimneys, Mach's "open book," or the reversible cubes. Thus, an intaglio can sometimes appear as a bas-relief, and vice versa. Figure 35 shows a simple example of an equivocal figure involving a group of rings or circles. Is the "slinky" receding toward the left or receding toward the right? A number of M. C. Escher's tantalizing graphics depend, for their effect, on embedded equivocal figures.

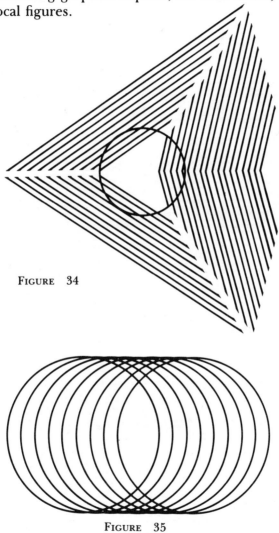

FIGURE   34

FIGURE   35

INDEX

References are to items, *not* to pages. A number followed by the letter *p* refers to the introductory material just preceding the item of the given number (thus 15*p* refers to the introductory material immediately preceding Item 15°).

INDEX